THE GAME OF SPECIES

THE GAME OF SPECIES

An Introduction to Biodiversity

JULIÁN SIMÓN LÓPEZ-VILLALTA

PELAGIC PUBLISHING

First published in 2025 by
Pelagic Publishing
20–22 Wenlock Road
London N1 7GU, UK

www.pelagicpublishing.com

The Game of Species: An Introduction to Biodiversity

Copyright © 2025 Julián Simón López-Villalta

The right of Julián Simón López-Villalta to be identified as the author
of this work has been asserted by him in accordance with
the UK Copyright, Design and Patents Act 1988.

All rights reserved. Apart from short excerpts for use in research or
for reviews, no part of this document may be printed or reproduced,
stored in a retrieval system, or transmitted in any form or by any means,
electronic, mechanical, photocopying, recording, now known or hereafter
invented or otherwise without prior permission from the publisher.

https://doi.org/10.53061/OSJH4854

A CIP record for this book is available from the British Library

ISBN 978-1-78427-558-7 Pbk
ISBN 978-1-78427-559-4 ePub
ISBN 978-1-78427-560-0 PDF

EU Authorised Representative: Easy Access System Europe – Mustamäe
tee 50, 10621 Tallinn, Estonia, gpsr.requests@easproject.com

Typeset in Stone Sans by S4Carlisle Publishing Services, Chennai, India

Cover image: Sand, Maurice, George Sand and Alphonse Depuiset,
1867. *Le Monde des papillons : promenade à travers champs*.
Paris: J. Rothschild. https://www.biodiversitylibrary.org

Printed and bound by Short Run Press Ltd, Exeter EX2 7LW

To all living beings that are indifferent to us

Contents

	Preface	viii
1	Introduction to the game	1
2	On the origin of pieces	11
3	The roles of the pieces	22
4	About the boards	33
5	The islands board	37
6	The Red Queen's board	46
7	Redux	58
8	The future of life	66
	Glossary	79
	References and further reading	84
	Illustration credits	96
	Index	99

Preface

Three things about life on Earth have always fascinated people: the very existence of living organisms, their enormous diversity, and their complexity. How did this kaleidoscope of shapes, colours and ways of surviving come about? Why are there so many species? Why do some groups of organisms have more species than others? Why are some places more biodiverse than others? Have there always been so many species? How are we causing species to go extinct, and how fast? What can we do to stop it? This book explores questions like these, in what I hope is a straightforward and accessible way.

Nature is not only a spectacle of breathtaking beauty, it is also a vast mystery, a labyrinth of questions that arise as we discover the wonders of life around us, questions that point the way to a better understanding of our natural heritage, and therefore to its appreciation and conservation. However, some naturalists react with a certain annoyance to the science that studies nature, as if a scientific approach to biodiversity, ecosystems and evolution would rob life of its magic. On the contrary, trying to understand life makes it even more magical. For, just as we cannot properly appreciate historical heritage without knowing something about history, we cannot properly appreciate biodiversity without some idea of what it is and how it works. In fact, it is dangerous to believe the opposite, because we run the risk of becoming the sorcerer's apprentice, in this case believing that it is enough to feel sympathy for nature in order to know how to protect it. This is not the case, just as it takes more than the intention to cure the patient to be a doctor.

In the preface to my first book, *El monte mediterráneo* (*The Mediterranean Shrubland*; Tundra Ediciones, 2016), I said that I had written it because it was what I would have wanted to read years ago when I started out as a naturalist. The same is true of the present book, but on a much deeper level. For this book, though shorter, is much broader in its approach. It is the fruit of hundreds of hours of wondering about the origins of biodiversity while observing nature, thinking about the same questions over and over again, searching for answers by diving into the scientific literature, consulting researchers or developing my own little research projects. As in my earlier book, I wrote these pages because as far as I could ascertain there was unfortunately no such book, in Spanish, English or any other language. Tundra Ediciones published a Spanish version in 2017 (entitled *El juego de las especies*), and now Pelagic Publishing has opened up the ideas contained in the book to many more potential readers with this English version, which is essentially the same book with some updates.

I wanted to write a book that would not put off anyone interested in biodiversity. It is important to know how species play the game of life on Earth, and so an understanding of the key rules of the game should reach as many people as possible. Sometimes it has been difficult to simplify what I wanted to say without sacrificing its complexity, but in the end I am happy with the middle ground I have found, and I would like these pages to be palatable to beginners and experienced naturalists alike. I have tried to avoid jargon and technicalities as much as possible, because they close many doors to those who approach nature with a desire to learn, and because if we can't explain the science of species in simple terms, then we probably either don't understand it, or what we have written doesn't really explain what we thought it did.

The main purpose of this book is to convey what I believe is essential to understanding how biodiversity works. In order to achieve this goal, it seemed sensible to leave out anything that I felt was not crucial or solidly proven. Thus food webs, coevolution, the unified neutral theory of biodiversity and biogeography, metabolic theory, punctuated equilibrium, species selection, macroecology and a host of comparative methods are omitted. They are simply not essential to what I want to say. Including them would only have added unnecessary complexity because, to paraphrase Michael Ende, I think they are part of another story and should be told another time. However, many of these topics are listed as glossary entries, for readers who wish to look them up.

It is also a different story when it comes to the world of microbial biodiversity, still largely unknown and often at odds with our usual idea of what a species is. If we take a species to be a group of organisms that can reproduce sexually among themselves, producing fertile offspring, how do we deal with bacteria and their strange relatives, the archaea, microorganisms that lack sexual reproduction but frequently exchange DNA despite belonging to very different groups in the tree of life on Earth? Could we then say that all these prokaryotic cells form a single species, or that their diversity is organised not into species but into distinct units? Whatever the answer, the fact remains that the biodiversity of bacteria and archaea is overwhelming, because of the unusual range of metabolisms they can develop that we do not find in the eukaryotes. While the true magnitude and diversity of this microcosm is still being unravelled, in the following chapters we will focus mainly on the diversity that is best known, that of plants, animals and other organisms in which species can generally be distinguished according to the above definition.

I would like to express my thanks to everyone who has contributed to the birth of this book. To the team at Pelagic Publishing for all their work and enthusiasm on this project, and to Hugh Brazier, whose editing gave the text the polish that biodiversity deserves. To all the authors from Wikimedia Commons whose images make these pages shine, for their generosity in making their work publicly available under the terms of the Creative Commons licences. I have tried to acknowledge them as much as possible, and not just in the credits section of the illustrations, sometimes going beyond what the licences require. I am also grateful to Ian Hutton, Deborah Kaspari and Alex Liu, who kindly provided me with illustrations for this project. To Ben Twist for his generosity in allowing me to use his photographs. To Peter Grant, Tiago Quental and Victor Savolainen for their thoughtful responses to my questions about their research. And especially to Michael Rosenzweig, for inspiring me with his legendary *Species Diversity in Space and Time* and for the wonderful months we spent thinking by email about the game of species in search of an ever-distant truth. Thanks also to my many beta readers, especially to Scully for her selfless editorial work – do you see how it didn't take eight years to write this one? Or wait, just a minute…

Campo de Calatrava, January 2025

Chapter 1
Introduction to the game

The game of species is the oldest game of all. It began billions of years ago, before the mountains rose. It is older than many stars. The game is played by microscopic algae on the ice of Antarctica, by trees as tall as skyscrapers on the Pacific coast of North America, and by extravagant creatures in the volcanoes of the deep sea. The game follows rules that, through a series of improbable events, have led to the evolution of those who discover them, of you and me, dear reader. It has created the living heritage that surrounds us, and continues to do so quietly, day after day.

If the word 'culture' has to do with 'cultivating', that is, making grow, then, in a very broad sense, the game of species is the most basic culture to which we belong, since it created us. However, few know the basics of how it works, apart from the ecologists and evolutionary biologists who study it. I am writing this book in an attempt to remedy that situation, and in these pages I would like to share the joys of getting closer to understanding this amazing game. Moreover, it is urgent that we understand it well. Without knowing the rules of the game, we will not be able to appreciate it properly or know how to preserve it in the midst of one of its greatest crises. *Homo sapiens*, one of the youngest pieces on the board, has taken over the whole game and is trying to destroy it, unaware that in doing so it is impoverishing itself and the entire world. We cannot let this happen, so we must use the rules wisely from within to turn the situation around.

What is the game of species? Nothing more or less than the functioning of biodiversity, from the struggle for life in a drop of water to what happens at the scale of the whole planet, from the fractions of a second it takes a chameleon to catch a grasshopper to the billions of years in which species come and go like fleeting images in the film of evolution.

Soon we will take a look at the great game of life on Earth, but first we need to equip ourselves with a few things for the journey. We need to know what biodiversity is: the variety of life at all levels, from genes to ecosystems, that is, biological diversity. A more detailed definition is given in the 2011 United Nations

Convention on Biological Diversity, which is arguably the most important global agreement for the conservation of biodiversity:

> 'Biological diversity' means the variability among living organisms from all sources including, inter alia, terrestrial, marine and other aquatic ecosystems and the ecological complexes of which they are part; this includes diversity within species, between species and of ecosystems.

In practice, the biodiversity of a place is often measured simply by counting the number of species present, which many refer to as species richness. Biodiversity can be measured in many other ways, using different indices, but in this book we will use the word 'diversity' to refer to the number of species. This is for two reasons: first, because of its simplicity, and second, because we know that it is usually highly correlated with the other measures.

There is nothing playful about the game of species for its players. It is a constant challenge of survival, won through a combination of luck and merit. But it is difficult to understand that the harshness of the game is not just a matter of words. The main objective of this chapter is to make it clear that in every square of the board of life there is a real background of ruthless indifference to living beings. Does that sound exaggerated? Let's see…

The struggle for existence

Genovesa is the name of a small island in the Galápagos archipelago. Here, from 1978 to 1988, the ornithologists Peter and Rosemary Grant followed the lives of most of the Cactus Finches *Geospiza scandens*, catching and tagging the birds, and building their family trees year after year. In that time, 79% of the breeding males and 78% of the breeding females failed to produce a single offspring that survived to reproduce. The most successful female of the 86 that the Grants followed bred eight times between 1978 and 1987, laying a total of 110 eggs. These produced only 58 chicks, of which only three survived to breed. These were not good times for the island's finches, but in nature, years of hardship come more often than the inhabitants would like.

Settled in an artificial world of houses, cars and supermarkets, we have forgotten how hard it is for other species to survive in nature. It seems to us that it must not be so difficult when we watch birds in the forest for a few minutes, because they seem to be living carefree in a beautiful and sonorous grove. But the truth is brutally different, and who better to tell it than Charles Darwin:

> We behold the face of nature bright with gladness, we often see superabundance of food; we do not see, or we forget, that the birds which are idly singing round us mostly live on insects or seed, and are thus constantly destroying life; or we forget how largely these songsters, or

FIGURE 1.1 The Cactus Finch *Geospiza scandens*, a bird unique to the Galápagos Islands. Left, a female; right, female and male. Painted in the 1830s by Elizabeth Gould to illustrate Darwin's book about his voyage around the world on HMS *Beagle*. The cactus finches of Genovesa have recently been assigned to a separate species, *G. propinqua*.

> their eggs, or their nestlings, are destroyed by birds and beasts of prey; we do not always bear in mind, that though food may be now superabundant, it is not so at all seasons of each recurring year.

In another part of *On the Origin of Species*, Darwin tells us something tremendous that he observed:

> on a piece of ground three feet long and two wide, dug and cleared, and where there could be no choking from other plants, I marked all the seedlings of our native weeds as they came up, and out of the 357 no less than 295 were destroyed, chiefly by slugs and insects.

He also noted the casualties caused by the elements:

> I estimated that the winter of 1854–55 destroyed four-fifths of the birds in my own grounds; and this is a tremendous destruction, when we remember that ten per cent. is an extraordinarily severe mortality from epidemics with man.

These and many other observations led Darwin to marvel at the myriad silent battles that animals and plants must fight to survive in nature:

> What a struggle between the several kinds of trees must here have gone on during long centuries, each annually scattering its seeds by the thousand; what war between insect and insect – between insects, snails, and other animals with birds and beasts of prey – all striving to increase,

and all feeding on each other or on the trees or their seeds and seedlings, or on the other plants which first clothed the ground and thus checked the growth of trees!

What is the root cause of all this destruction? First and foremost, the fact that more living beings are born than can survive with the resources available in their environment. For example, a tree can easily produce thousands of seeds over a lifespan of several centuries, but when it dies in the forest, its place is taken by a single tree of its species, born from just one of the seeds of all the trees in the area. In other words, an astronomical number of seeds fail. The alternative to this is to ask for the impossible: a finite forest with an infinite number of trees, or to generalise, a finite world with an infinite number of living things. We can imagine the situation of any other wild creature, and the result will be the same: many are born, but few are chosen to survive and reproduce, perhaps by chance alone. Even an animal as slow to reproduce as the elephant is no exception to this rule: Darwin calculated that a single pair could produce about 19 million offspring after only 740–750 years. We do not observe this spectacular progression in the number of living beings simply because nature does not allow it.

Famine, the vagaries of the weather and innumerable enemies are constantly holding back the growth of life. In Darwin's words: 'A struggle for existence inevitably follows from the high rate at which all organic beings tend to increase'. Darwin came up with this idea when he read the economist Thomas Malthus, who predicted that humanity would face serious food supply problems because its population was growing faster than the world's food production. Darwin noted that this was similar to the general situation in nature:

> Hence, as more individuals are produced that can possibly survive, there must in every case be a struggle for existence, either one individual with another of the same species, or with the individuals of distinct species, or with the physical conditions of life. It is the doctrine of Malthus applied with manifold force to the whole animal and vegetable kingdoms; for in this case there can be no artificial increase of food, and no prudential restraint from marriage.

Hence comes his expression 'struggle for existence', which, Darwin himself wrote, should be understood in a broad sense, and not literally. He also warned us that it is very difficult to keep this struggle in mind at all times when we think about nature. And we should, because the struggle for existence is the basis for understanding the interplay of species, as the following stories will illustrate.

House Sparrows, natural selection and bottlenecks

In the early hours of 1 February 1898, a terrible snowstorm hit the city of Providence, Rhode Island. The next morning, someone picked up 136 dying

House Sparrows *Passer domesticus* and took them to the Anatomy Laboratory at Brown University. Within a few hours, 64 had died and 72 had reanimated. Hermon Bumpus, a professor of zoology, realised that he had an excellent opportunity to test natural selection. He took several measurements of each bird, and found that the survivors were on average longer and heavier, with a smaller and narrower head, and shorter keel and legs. In short, they were bigger and more compact than those that died. Because of these anatomical differences, the sparrows that survived were less susceptible to chilling. Exactly what we would expect by natural selection. So Hermon Bumpus not only recorded an extreme case of the struggle for existence against the elements, he also gave us what is often regarded as the first direct evidence of natural selection.

Natural selection occurs whenever certain hereditary characteristics provide advantages or disadvantages for survival, or reproduction, or both. As a result of natural selection, from generation to generation, the advantageous hereditary characteristics (adaptations) become more widespread and the disadvantageous ones gradually disappear. This process changes the heritable characteristics of a population over time. The result is that the population evolves.

But there is also evolution without natural selection. For every accident that changes the DNA of an individual (mutation), there is a genetic change in the population, that is, evolution. And when immigrants arrive (migration), they can bring new genes from outside into the population, or old ones but in different proportions. Another way of evolving without natural selection is called genetic drift, and it explains why about 10% of the human inhabitants of the island of Pingelap, in the Pacific Ocean, cannot see colours (achromatopsia). In 1775, a devastating typhoon left about 20 survivors on this small atoll, and their chief, Nanmwarki Mwanenihsed, must have carried the recessive gene that causes this visual condition. Tribal inbreeding did the rest, and so we have this example of evolution by 'bottleneck', or the 'founder effect'. A founder effect occurs when an event leaves so few survivors that any of the genes they happen to carry can become common in the offspring, just by chance, even if that gene is not useful. All these evolutionary mechanisms exist and are very interesting, but for now let's focus on natural selection, because it is a crucial key to the game.

Peppered Moths and industrial melanism

Perhaps the most famous and exhaustive proof of natural selection is the case of the Peppered Moth *Biston betularia*. In England, until the early nineteenth century, this moth was mostly light-coloured with dark speckles, a form known as *typica*. In 1848, an almost black, melanic version, the *carbonaria* form, was first found. In the years that followed, these dark moths became increasingly common, to the detriment of the light ones. Thus, by 1895, 98% of Peppered Moths were dark. It was around this time that the entomologist James William

FIGURE 1.2 The Peppered Moth *Biston betularia* is an excellent example of natural selection. The light-coloured *typica* form (left) is well camouflaged on light trunks, and is thus protected from its enemies the birds. The *carbonaria* form (right), on the other hand, is less conspicuous on trunks in polluted areas, because the bark is obscured by soot from the smoke. Thus, as industrial pollution spread in the nineteenth century, the dark form increased to the detriment of the light form. Photos by Donald Hobern (left) and Jerzy Strzelecki (right).

Tutt suggested that this change was due to natural selection. He noted that the industrialisation of England had brought intense pollution, including sooty smoke which changed the predominant colour of the tree trunks in many places from light to dark. When resting on these dark trunks during the day, melanic moths were much better camouflaged than light moths. For this reason, he argued, insectivorous birds found the light moths easier to spot and therefore ate them more often than the dark ones. This process equates to natural selection against the light form and in favour of the dark moths.

For decades, no one bothered to confirm whether Tutt was right, as it seemed doubtful to many that these moths, when resting on tree trunks, could be frequently eaten by birds. But in the 1950s the entomologist Bernard Kettlewell carried out a series of experiments, with hundreds of Peppered Moths, the results of which confirmed Tutt's supposition. As a result of these experiments, the industrial melanism of *Biston betularia* began to be seen as a milestone in the study of evolution, and became a textbook example for explaining natural selection. Its genetic basis was discovered, with melanism found to be determined by a single gene, and other cases of industrial melanism were found in several butterfly species.

However, at the beginning of the twenty-first century the validity of Kettlewell's experiments came under serious attack from creationists, and also from some biologists. The most reasonable doubts concerned details of the experimental design, and the importance of those doubts was exaggerated by the most critical opponents. In the scientific arena, this led to a bitter debate between supporters and detractors, even to the point of personal insults. Creationism and the most sensationalist press took advantage of the troubled waters to portray the entire theory of evolution as a fraud. Spurred on by this situation, Michael Majerus, an entomologist and staunch adversary of creationism, conducted new, more carefully designed experiments, based on observations of 4,864 Peppered Moths over a seven-year period. The results were published in 2012, after Majerus' death, and laid to rest all previous legitimate criticisms. Majerus had succeeded in documenting natural selection at work in *Biston betularia*, and the story was just as Tutt had assumed more than a century earlier.

Salmon and sexual selection

Contrary to the claims of creationists, there are many precise and meticulous verifications of evolution by natural selection. Some of these cases of natural selection have to do with resisting the elements of climate, as we saw with Hermon Bumpus's sparrows. Meteorological vagaries can intensify competition for a sheltered site, or a humid or dry one, or for food when it is scarce. Other examples of natural selection concern defences against predators, such as that of the Peppered Moth. Still others involve competition through various characteristics. For example, in order to reproduce, Coho Salmon *Oncorhynchus kisutch* return from the northern Pacific Ocean to the rivers where they were born, in

FIGURE 1.3 Above, a male Coho Salmon *Oncorhynchus kisutch*. Below, Guppies *Poecilia reticulata* of the wild variety, one male and two females. Photo by Per Harald Olsen.

8   THE GAME OF SPECIES

Russia, Alaska and other regions. In their natal riverbeds, females make up to seven 'nests', depressions where they lay their eggs, which are fertilised by one or more males. The females then cover the fertilised eggs with gravel and defend their nests for an average of eight days, until they die. The size of females has

FIGURE 1.4 Female and male Red Bird of Paradise *Paradisaea rubra*, from Indonesia. The splendid finery of the male is a good example of the oddities that sexual selection can create, for this showy plumage serves only to attract the female, and for that reason alone it has evolved, favoured by natural selection even though it can be detrimental to the male in other ways – for example, by making him more visible to predators. Illustration by the zoologist Daniel Giraud Elliot for his 1873 book *A Monograph of the Paradiseidae or Birds of Paradise*.

been shown to be subject to natural selection, as they have up to 23 times more reproductive success if they are larger. This is because size helps them to produce more eggs, to gain a riverbed with better conditions for egg development, and to defend the nests effectively.

It is difficult to think of a characteristic that cannot be subject to natural selection. Sexual attractiveness often is, and we have many examples; Darwin called this kind of selection 'sexual selection'. Due to sexual selection, certain characteristics of one sex spread through the population simply because they are more attractive to the opposite sex, for whatever reason. This is the only way to explain the spectacular ornaments and quirky dances of the male birds of paradise. It is not uncommon to discover that each sex has good reasons for preferring the seemingly odd traits it seeks in the other. In birds, for example, colourful and bright plumage usually indicates a healthy individual with few parasites, and what better parent for your offspring than such a bird? It is logical that natural selection would favour this taste in the female, and therefore the development in the male of a beautiful plumage, which clearly shows 'what one is worth'.

Guppies: rapid evolution

Examples like these, from the sparrows to the birds of paradise, show us that natural selection exists and is capable of causing marvellous changes in living beings. Furthermore, it can act very quickly; contrary to popular belief, evolution does not require millions of years to happen. A good example of this stars a colourful little fish from tropical rivers of America, the Guppy *Poecilia reticulata*, well known among warm-water aquarium hobbyists. In Trinidad, Guppies can be preyed upon by fish from the cichlid and characin groups, some of which preferentially select large victims. The Guppies that coexist with these enemies have evolved to cope with the high risk of being hunted as they increase in size. Among other things, they reach reproductive maturity at a younger age and smaller size than Guppies from rivers lacking these predators. Since they mature young, they can breed before the likely fatal encounter with their voracious neighbours.

Into the small world of these fish came the biologist David Reznick and his coworkers, to carry out a field experiment. They located a stream with a waterfall, below which lived the Guppies and their enemies, while upstream of the waterfall there were neither Guppies nor predatory fish (except for the small *Rivularia hartii*, an omnivore that sometimes hunts an occasional Guppy). The scientists took Guppies from downstream and introduced them above the waterfall, thus releasing them into a much safer environment. They repeated the same experiment in another stream. After 11 and 7.5 years, respectively, the scientists returned and found that the introduced Guppies had evolved: they now matured later, were larger than before, and had other traits typical of

Guppies from streams with few predators. What had happened? Above the waterfall, natural selection had favoured late-maturing Guppies, because their larger size meant they could reproduce more successfully (like the Coho Salmon). Reznick and his team bred these Guppies in captivity, and so they were able to confirm that the observed changes were indeed heritable. So in just a few years, natural selection made these fish evolve quite a lot.

All these stories testify to the idea that the game of species is played on a board which is very often harsh and hostile, where the struggle for survival and reproduction is commonplace, and where the resulting natural selection can cause important evolutionary changes even within a few years. So Darwin was right. But is his other big idea, that natural selection produces new species, also true? Let's look at this in the next chapter, which will help us understand the origin of the pieces that play the most wonderful game of all.

Chapter 2

On the origin of pieces

In the game of species, the variety of pieces is absolutely prodigious. Few people truly understand how full of species nature is; few imagine that, in temperate latitudes, a small hill with some wild vegetation can easily harbour hundreds of species of plants and animals. As of this writing, over two million species have

1. Geospiza magnirostris.
2. Geospiza fortis.
3. Geospiza parvula.
4. Certhidea olivasea.

FIGURE 2.1 When Darwin discovered several species of finches unique to the Galápagos Islands, he thought it was as if a single species had been modified to feed in different ways, developing thicker beaks for cracking seeds, or thinner ones for catching insects, but maintaining a family resemblance. From these and many other observations, he concluded that natural selection changes species so that new ones originate. This iconic illustration of Darwin's finches was created by John Gould in around the 1830s.

been described, of which more than half are insects – the game seems to be played largely with six-legged pieces. The best known groups, plants and vertebrates, contribute about 400,000 and 75,000 species, respectively. And we know that life on Earth still has many, many more species waiting to be discovered, especially insects, up to a total of perhaps five million, maybe even eight million.

Why are there so many species? Why isn't there, for example, a single 'higher' species that has extinguished all the others? How does a species originate? Let's look for answers by delving into the 'mystery of mysteries', as Darwin called the origin of species. In his day that expression was well justified, for the fog of ignorance obscured the subject, but thanks to his work, and that of many others, we now know a great deal about how species arise, that is, about speciation.

The central idea proposed by Darwin in On the Origin of Species is that natural selection gradually modifies species, forming varieties within them that change until eventually one becomes so different that we consider it a new species. In general, this is true, as we will see in this chapter. Some people still reject the idea that natural selection can form new species, but evolutionary biologists no longer argue about this because an abundance of proofs closed that debate decades ago. Nevertheless, the task of documenting all the details of the speciation process in nature is so challenging that we do not have many examples with all the loose ends completely tied up.

Natural selection results in new species

In the light of all the evidence, it seems that most species form by a mechanism called ecological speciation: a population adapts to particular conditions through natural selection, changing until it becomes so different from the rest of its species that it no longer breeds with them, or very rarely, even if they coexist. When the evolving population reaches this point, it has become a new species, because it shows what we consider the key to being a species: reproductive isolation from other similar species, at least to a large extent.

This is the moment to comment a little more on the concept of species, or, in the terminology of this book, to think about what a 'game piece' is. Most biologists would agree that a species consists of individuals that can reproduce with each other, at least potentially, and produce fertile offspring. This is the most common concept of species, the *biological species concept*. But there are other species concepts, which define a species as a group of organisms that share the same traits (*morphological species concept*, useful for fossils), or as a single branch in the tree of life (*lineage* or *phylogenetic species concept*, which relies heavily on genetic data). The lineage species concept can be applied even to asexual organisms, where the biological species concept is impossible, although in asexuals it is difficult to objectively set the threshold of genetic difference for defining a species.

In practice, evolution tends to reconcile these three species concepts by making organisms that differ in all three respects: they can no longer successfully reproduce with each other, they have different morphologies, and they have different genetic makeups (or genotypes). Thus the biological species concept, although not without nuances and shortcomings, provides a solid basis for understanding the main lines of the game of species, in sexual organisms at least. From this standpoint, let's continue with our question: how does natural selection form species?

An impressive proof that species arise by natural selection, as Darwin proposed, is parallel speciation, in which new species arise in many independent cases from the same ancestor, when it is subject to the same ecological pressures, resulting in the evolution of the same traits. This is best understood with an example, and a useful one is that of a small boreal fish, the Three-spined Stickleback *Gasterosteus aculeatus*. These fish are widely distributed in North America, Europe and Asia, but we are particularly interested in the sticklebacks that live in the coastal lakes of British Columbia, Canada, which formed as the glaciers retreated some 13,000 years ago. Many of these lakes are inhabited by two types of stickleback, which do not reproduce with each other – and so, according to the biological species concept, they would be distinct species. Invariably, one type is adapted to forage on the bottom near the shore, and the other, smaller and more slender, swims near the surface and catches a lot of plankton. These

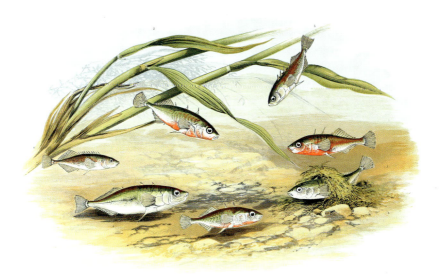

FIGURE 2.2 Three-spined Sticklebacks *Gasterosteus aculeatus*. Reproductive males are distinguished by their reddish hue in the ventral parts. They build and defend a nest into which females are invited to enter to lay eggs, which the male will fertilise later. Watercolour by Alexander Francis Lydon, published in W. Houghton's *British Fresh-Water Fishes* (1879).

two types (species) are presumed to have evolved independently, in parallel, in several different lakes. In some lakes, however, there is only one type of stickleback, with intermediate characteristics, an 'all-terrain' fish that feeds both on the bottom and on surface plankton. These are the protagonists of our first story of parallel speciation in action.

Experimental proof of the concept of parallel speciation in sticklebacks came with an innovative experiment by evolutionary biologist Dolph Schluter. He chose a lake with only 'all-terrain' sticklebacks, introduced surface-adapted sticklebacks, and waited. After a single generation, the effects of natural selection were overwhelming. The resident all-terrain sticklebacks whose body proportions were similar to the new surface sticklebacks grew small and scrawny, while those more similar to bottom sticklebacks grew well and became much fatter. Clearly, in the struggle for survival, the surface sticklebacks were beating the intermediate sticklebacks that more closely resembled them. And so, with the blind determination of a physical law, natural selection was creating a bottom type from the intermediate type. This suggests that, in many different lakes, competition between sticklebacks had triggered natural selection to modify the intermediate type until two new species formed, one for the bottom and one for the surface. The lack (as yet) of a scientific name for these *Gasterosteus* species, and for the many more that have formed in these lakes, does not detract from the fact that they meet the essential criteria to be considered distinct species.

Another story of parallel speciation involves much more sinister actors: lampreys. Many of these jawless fish are parasites that suck the blood of other fish, and live in the sea but return to the streams where they were born to reproduce. Some of these streams are home to another species of lamprey, usually small, which does not descend to the sea. It is called a 'satellite species' because it has evolved from the parasitic lamprey by adapting to the stream environment. Among other adaptations, the satellite has ceased to be parasitic, since in streams parasitism is not a good option for a lamprey (given the lack of large fish to parasitise, or their scarcity in certain seasons). In Europe, the Brook Lamprey *Lampetra planeri* is a satellite species evolved from the European River Lamprey *L. fluviatilis*, and genetic studies have shown they are distinct but very closely related species. The origin of satellite lampreys has happened time and again, on different continents, and always, it seems, because some species of marine lamprey adapted to life in streams. How else can we explain this repeated speciation, if not because natural selection has acted on marine lampreys by adapting them to streams again and again, on different continents? I think Darwin would have been delighted to find examples like this, as direct evidence for the origin of species by natural selection.

Moreover, Darwin would have liked to see that it is not only species that can form in parallel, but also varieties within a species (also known as forms, morphs or ecotypes) – as would be expected if natural selection works as he thought it

FIGURE 2.3 With a mouth like a suction cup, lampreys belong to a branch of vertebrates that pre-dates the evolution of jaws. From top to bottom: Sea Lamprey *Petromyzon marinus*; European River Lamprey *Lampetra fluviatilis*, adult; two Brook Lampreys *L. planeri*; and a young European River Lamprey. Illustration by Alexander Francis Lydon, published in W. Houghton's *British Fresh-Water Fishes* (1879).

would. Because, before forming a new species, Darwin supposed that natural selection would form a variety, so that if there is parallel speciation there should also be parallel production of varieties.

Indeed, that is just what we see in nature, and a good example is given by the Rough Periwinkle *Littorina saxatilis*, a sea snail common along the rocky coasts of the North Atlantic. In Galicia, England and Sweden, natural selection seems to have divided the species into two forms that coexist within a few metres of each other: a small one, adapted to cling well to rocks (with small shell and large foot), and a larger one, adapted to withstand the impacts of pebbles and to better resist its enemies the crabs (protected by its large and robust shell with a small opening). These two forms do not reproduce with each other very often, as if they were species in the process of formation. And this duo of 'quasi-species' has evolved again and again, independently on different coasts, in areas separated sometimes by only a few tens of kilometres, as genetic studies have shown.

The evolution of varieties by natural selection has also been confirmed by experiment. One such experimental study focused on stick-insects of the genus *Timema*. If we move these insects from the bushes where they usually live to different bushes, their camouflage will not work as well, because it is tuned specifically to resemble a certain type of foliage. Thus, in the new bushes, predators will see them more easily and eat the ones that stand out the most. In this way, natural selection will favour the insects that look more like the new foliage. Therefore, in a short time, we will have stick-insects of a different colour to the ones we started with, and we can observe that the offspring will inherit this difference. We have created a new variety of stick-insect! This is exactly what happened in the experiments, and it was also found that the new variety does not reproduce well with the original. How long will it take before the new variety becomes a new species?

These natural histories demonstrate how important natural selection is in the origin of species. In most known cases of speciation, selection of ecological traits appears to be crucial, with sexual selection secondary. Although the formation of a new species involves disrupting sexual reproduction with the ancestor species, sexual selection is generally considered to play a minor role in speciation as a whole. However, it can be critical for some organisms, as we know is the case for the cichlid fish of the African Great Lakes Victoria, Tanganyika and Malawi. These colourful fish, with a second pair of jaws in the throat, have often produced species simply because some females prefer to mate with red males and others with blue males. In many cases, these whims have divided a group of fish into two distinct species. However, in the cichlids the role of ecology has also been fundamental in speciation, as evidenced by their very varied lifestyles, which even include specialists in eating scales, which they remove from other fish.

Geographic Isolation

In the examples above, the origin of new species does not require geographic separation, but such isolation greatly facilitates the process; in fact, to form species, evolution usually needs to take place in geographic isolation. Otherwise, the changes caused by natural selection can be lost almost as quickly as they arise. If a region is not isolated from others (for example, by a sea or a mountain range), then migrants could arrive from areas subject to different conditions of natural selection. If these newcomers interbreed with the indigenous population, the resulting genetic mix will easily undo the effects of natural selection in that region.

Such genetic homogenisation is practically impossible on remote islands. Therefore, continental species that happen to arrive on these islands will easily give rise to new species. This helps to explain why archipelagos are so rich in unique species, in endemics that have evolved in situ. Take the Hawaiian Islands as an example: 89% of the flowering plant species in this archipelago are endemics, a very high percentage, similar to that of the island of Madagascar. Even on the tiny island of Lord Howe, 580 kilometres east of Australia, almost half of the plant species are endemics, even though the whole island measures just 10 × 2 kilometres.

We also find many endemic species in so-called island habitats, which resemble an island because they have very different conditions compared to the surrounding area. For instance, high mountain peaks are usually 'islands' of rock, cold and wind, immersed in a much less hostile 'ocean'. Due to their peculiar conditions and isolation, such island habitats are breeding grounds for new species. This explains why, in most mountain ranges, as we ascend there is an increasing proportion of endemic species. This is the case in the Sierra Nevada of southern Spain, on whose high summits the percentage of plant species endemic to this mountain range is around 30–40%, reaching as high as 80% on the stony outcrops of the very tops. This is a number more typical of islands than of continental habitats.

Lakes which are very large and very old constitute another island habitat, curiously being just the opposite of an island – a waterbody surrounded by dry land. Here we see a repeat of the previous story: the African Great Lakes are home to hundreds of species of cichlid fish, of which at least 97% are endemic to each lake. The percentage of endemics exceeds 99% for the water fleas (amphipods) of Lake Baikal in Siberia, where many other groups of invertebrates show similarly high endemicity. The great age of these lakes, and several others elsewhere in the world, has given evolution time to produce many species, sometimes at a very rapid pace.

FIGURE 2.4 The Blue-breasted Fairywren *Malurus pulcherrimus*, a small insectivorous bird, is found only in southwestern Australia, the island-continent where fairywrens evolved. There is evidence that the many different fairywren species originated by geographic isolation in different corners of Australia (allopatric speciation). Left, male; right, female. Photograph by Ben Twist.

18   THE GAME OF SPECIES

FIGURE 2.5 On Lord Howe Island, east of Australia (above, classic view from the north), two species of palms have evolved (below, foreground): *Howea belmoreana* (left, with whitish leaves) and *H. forsteriana* (right, much more slender). Photographs courtesy of Ian Hutton, Lord Howe Island Museum.

These examples, and many others, strongly suggest that geographic isolation has been key to the origin of many species. The speciation associated with geographic isolation is called allopatric speciation (allopatric = 'a different homeland'). The predominance of allopatric speciation as the main mechanism producing species is obvious when we take a look at any field guide with distribution maps on a continental scale, since we will often find genera whose species inhabit separate but adjacent regions. The most likely explanation for this is that these species arose through geographical isolation in distinct regions.

All these observations support the widespread idea that allopatric speciation is the main driver of biodiversity. But this does not mean species cannot originate in other ways. The alternative mechanism, sympatric speciation (sympatric = 'the same homeland'), does not require geographic isolation.

Sympatric speciation has been for decades the 'ugly duckling' of evolution. By the mid-twentieth century, many evolutionary specialists dismissed it as unlikely, arguing that, within the range of a species, genetic mixing caused by internal crossbreeding must generally be too intense for a new species to form by natural selection. But in recent decades new theoretical models and data have emerged that have changed this situation, and now the ugly duckling has become a swan, which only means that today most biologists consider that sympatric speciation can occur in nature, even if it is not the norm. For example, the sticklebacks at the beginning of this chapter are an excellent example of sympatric speciation, since two species originated in the same lake.

Another well-documented case involves the palm trees of Lord Howe Island, mentioned above. On this volcanic island, about seven million years old, two endemic species of palms of the genus *Howea* arose about two million years ago, as estimated from their genetic differences. It is hard to believe that there was any geographical separation between them, since the island covers barely 14 square kilometres and palm pollen spreads with the wind, allowing the trees to reproduce with each other over long distances. The fact that there are two distinct species of *Howea* palms shows us, as clearly as possible, that sympatric speciation does indeed sometimes occur in nature.

Instant species

It is not science fiction – species can also be born in an instant! It happens frequently in plants through a kind of sympatric speciation mediated by a genetic accident: the multiplication of the number of chromosomes in a single generation. This gives rise to an organism known as a polyploid.

Polyploids are the result of two genetic labyrinths. Sometimes they are formed by the joining of two sex cells that carry two sets of chromosomes instead of just one, due to an error in cell division. This will produce an organism with twice as many chromosomes as normal. If this polyploid survived (in animals this is very difficult) and reproduced with the species from which it came, it would produce an offspring incapable of reproducing sexually, as it would have three versions of each chromosome and this number would prevent cells from forming viable sex cells (because meiosis separates pairs of chromosomes, not threes). Such a situation means that the polyploid is actually a new species, because in practice it can no longer mix its genes with its original species. Such polyploids, arising from a single species, are called autopolyploids, and they are very unusual, as it is rare for an egg and sperm to form with a double chromosome number and even rarer for them to meet and fuse. Much more common, allopolyploids are born when two different species cross and the chromosome number doubles in the resulting hybrid by accident. An allopolyploid also becomes a different species from its parent species, for the same reason as an autopolyploid.

But a polyploid, of whatever kind, is not an evolutionary dead end, for it can reproduce sexually if it finds others like itself, perhaps born of itself by asexual reproduction (not unusual in plants). And polyploids can give rise to new polyploid species, each with an increasing number of chromosomes. If the basic chromosome number of a species is called n, and a eukaryotic organism is usually diploid, that is, $2n$, a first-generation polyploid would be a tetraploid ($4n$), which could in turn produce an octaploid ($8n$), and so on. If the origin of a tetraploid is unlikely, the origin of an octaploid is even more improbable, but in spite of this even $38n$ species have been identified in nature. Over time, since polyploids do not interbreed with their ancestor species, they accumulate evolutionary changes and become very distinct.

These genetic oddities are more common than we might think: polyploidisation accounts for 15% of known speciation cases in flowering plants, and 31% in ferns. For some reason that remains unknown, the ratio of polyploid species in plants tends to increase from the tropics towards the poles; in a suggestive similarity, it also increases from the warm lowlands to the cold mountain tops. Some cases of polyploidisation are known in animals, but they are much less frequent than in plants. This may be because, in plants, asexual reproduction is so common that a single polyploid will often be able to start its own lineage and thus avoid extinction.

Since polyploidisation leaves a clear mark on chromosome numbers, it is easy to detect in the resulting species. Because of this, we can be sure that it is not the most common mechanism of speciation, but it is the fastest, as it requires only one birth. This is why some have called it 'instantaneous speciation', or 'quantum speciation'.

In some cases, a new so-called species arises from the crossing of two species but without polyploidisation, that is, there is no multiplication of chromosome number. Imagine that two closely related species coincide in the same region and here they interbreed more or less regularly, producing hybrids that are not polyploids, but which we consider to be a new species (should we?). This speciation by hybridogenesis has happened, for example, in the Italian Sparrow *Passer italiae*, which evolved from the crossing of the House Sparrow *P. domesticus* and the Spanish Sparrow *P. hispaniolensis*. Cases like this are unusual, but more common than we previously thought. However, the fact that two species can produce viable hybrids with some regularity casts doubt on their status as truly distinct species. Are they quasi-species? More than almost-species? To add to the puzzle, the Italian Sparrow hybridises with the House Sparrow, but there is no evidence of hybridisation with the Spanish Sparrow when they coexist. With cases like this, nature seems to defy our mania to pigeonhole everything into different words.

***

Broadly speaking, we think the pieces of this amazing game originate as this chapter has described. Its purpose has been to illustrate in particular two fundamental points about the origin of species: (1) speciation is often driven by natural selection, and (2) geographic isolation greatly facilitates speciation and is usually the driving force behind biodiversity. At the end of the book, these two points will be useful in addressing big questions about biodiversity, such as why there are so many species in some areas and so few in others. Meanwhile, in the next chapter we will continue to build our game manual by looking at how species manage to survive without driving other species to extinction.

Chapter 3

The roles of the pieces

The game of species depends largely on the role that each species plays in nature. This role is the species' ecological niche. Or think of nature as the performance of a play in the theatre; each species is a character in the play, and the role it plays is its niche. In reality, the niche is the set of relationships that a species establishes with the environment and with other species. This sounds terribly vague and abstract, and unfortunately it is ... One of the big problems of ecology is that we still lack a simple and objective way of identifying and measuring niches. As a remedy to this situation, the ecologist Eric Pianka proposed the creation of 'periodic tables of niches', but the idea has not been very popular.

Moreover, there is no single, universally accepted definition of the ecological niche. The term 'niche' was coined by Roswell Hill Johnson in 1910, but the idea had existed long before that. In the nineteenth century, naturalists already thought that you have to look at what species do to understand the plot of the whole play. That is why Darwin wrote about the 'positions in the economy of nature' occupied by species – in modern terms, he was referring to ecological niches. Later, at the beginning of the twentieth century, several ecologists came up with their own versions of the niche. For Joseph Grinnell (1917), the niche is

TABLE 3.1 A simplified periodic table of niches, for a Mediterranean shrubland. In each box, the organisms on the left exploit their environment mainly in two dimensions, and those on the right do so in three dimensions because they climb well or fly.

Size	Plants	Herbivores		Carnivores	
small	annual plants	caterpillars grasshoppers aphids	bees beetles butterflies	mantises wolf spider scorpion	wasps robberflies lacewings
medium	small shrubs	voles mice	dormouse goldfinch	lizards hedgehog	blackbird shrike
big	tall shrubs	rabbit hare	wood pigeon	Egyptian mongoose fox	genet imperial eagle
very big	trees	wild boar deer		Iberian lynx wolf	

the way in which an animal uses a given habitat, that is, the sum of the habitat where it lives and how it uses it. But for Charles Elton (1927), the niche of an animal would be its position within the food chains, or, in his own words, 'its relations of food and enemies'.

The most sophisticated view of the ecological niche was proposed in 1957 by the ecologist George Evelyn Hutchinson, for whom the niche of a species is nothing less than an *n*-dimensional hypervolume. This requires the following explanation: consider, for example, how the abundance of a grass species changes as soil moisture varies. If we plot abundance against moisture, we will get a graph that we can represent as a line, straight or curved. If we add another axis to the graph, say the temperature factor, the line will become a flat, two-dimensional object. A third ecological factor (it could be the abundance of herbivores) will turn it into a three-dimensional object, a volume. More than three ecological factors will give us what is called a hypervolume, an abstract object with as many dimensions as the factors we have included. Hence the niche is mathematically an *n*-dimensional entity, where *n* is the number of factors considered.

Each piece to its post

Now we must ask a key question: how can we combine the idea of ecological niche with that of natural selection? If two species have very similar niches, and they coexist, they will probably compete with each other extensively and often. In the face of such competition, natural selection tends to separate the niches, so that the two species are spared much of the disadvantage of competing, and both benefit. This separation of ecological roles is called *niche differentiation*, or *niche segregation* – when we refer to the change in the characteristics of the corresponding species, then we speak of *character displacement*.

There are some incredibly detailed examples of this niche shifting due to competition. In Chapter 2 we explored one of the best, the story of the sticklebacks. In that example we learned about an experiment which showed that, when two species of sticklebacks competed (an intermediate type and a surface-feeding type), the intermediate type underwent tremendous natural selection towards traits that were the opposite of those of the surface type, drifting towards a bottom-feeding type. In this way, the niches of the competing sticklebacks were separated: the surface type continued to use its preferred part of the lake, and the intermediate type was 'invited' by natural selection to leave this part of the lake free for its competitor, and to devote more of its time to the bottom.

Another extraordinary example of niche differentiation involves the famous Darwin's finches of the Galápagos Islands. For many years, the only ground finch found on Daphne Major, a small volcanic islet consisting almost entirely of a single crater, was the Medium Ground Finch *Geospiza fortis*. But in 1982,

FIGURE 3.1 Two granivorous birds of the Galápagos Islands: the Large Ground Finch *Geospiza magnirostris* (left) and the Medium Ground Finch *G. fortis* (right). The two species sometimes compete for the fruits of *Tribulus cistoides*, such as the one the Large Ground Finch is cracking in this picture. During a famine for these birds on the island of Daphne Major, such competition triggered rapid natural selection, which, within a few years, led to the Medium Ground Finch becoming smaller-beaked on average. Illustrations by the author.

the Large Ground Finch *G. magnirostris*, a potential competitor, began to breed on the island. This larger finch cracks many fruits of a plant known as Abrojo (*Tribulus cistoides*) to eat the seeds. It is able to do this thanks to its robust bill, which is much thicker and larger than that of the other species. Some Medium Ground Finches also occasionally eat *Tribulus* seeds, but they find them harder to crack than the Large Ground Finches, who quickly chase the Medium Ground Finches away from the tasty seeds.

It happened that two very dry years came, 2003 and 2004, so bad that no finch was able to breed on Daphne, and many died of starvation. Faced with this famine, the Large Ground Finches made the most of the few *Tribulus* seeds that were available, leaving almost none for the Medium Ground Finches with thicker beaks, which could have made use of this resource. Meanwhile, the Medium Ground Finches with thinner beaks probably survived on the seeds that were easier for them to eat, such as those of prickly pears. In other words, in these famine years, natural selection favoured the Medium Ground Finches with thinner bills and worked against the thick-billed individuals.

What was the result? That by 2004 and 2005 the few remaining Medium Ground Finches on Daphne Major had, on average, much thinner beaks than a few years earlier. And so the two finches separated their niches, with the Large Ground Finches remaining as thick-billed, consummate *Tribulus* eaters and the Medium Ground Finches now having thinner beaks, better suited to other food sources. All the details of this amazing story are based on the extensive fieldwork carried out in Galápagos by the ornithologists Peter and Rosemary Grant. As Peter Grant kindly informed me (January 2017), on this islet the Medium Ground Finch has remained thinner-billed ever since. In other words, the difference caused by natural selection is still there, even though many years have passed. Perhaps the most striking aspect of this example is that competition can take decades to occur, but when it does it can have dramatic and long-lasting effects in a short time.

Apart from these two excellently documented cases, there are many, many others which appear to be character displacement, and which, if studied in depth, could certainly reach the same level of detail as the cases outlined above. For instance, two North American species of lungless salamanders, *Plethodon cinereus* and *P. hoffmanni*, are very similar when living apart, but where they coexist they show differences, with the former becoming smaller and hunting smaller prey, and the latter vice versa.

It is clear that when niche segregation occurs, the niche becomes narrower, at least for one species. As its niche contracts, the species becomes more specialised. This increased specialisation, this creation of specialists, is the main consequence of natural selection when it acts on niches. This is not just a curiosity; it seems to be the cornerstone of how communities of life are organised in nature. When we get to know an ecosystem in depth, we almost always find that species organise themselves by exploiting resources from different niches. Sometimes the differences between their niches are very subtle, but they are there if we look carefully. For example, in the 1950s the ecologist Robert MacArthur made meticulous observations of five North American species of small birds (warblers of the genus *Setophaga*, formerly *Dendroica*). He saw them foraging for food on the branches and trunks of the trees where they coexisted. Over the course of hundreds of observations, he noted where in the tree each bird foraged and for how many seconds. He found that each species tended to hunt insects in a fairly specific area of the tree, thus avoiding too much competition with the other species. He did not show that this niche differentiation was caused by natural selection, but he found the partitioning of resources that is probably the footprint of that process. And there are many, many other similar examples.

A lethal competitor

What happens if a species does occupy a niche that coincides with those of other species? It could even disappear from the ecosystem. This sounds reasonable, but we also have experimental proof, especially since the Russian microbiologist Georgii Frantsevich Gause carried out a series of experiments with microbes in the 1930s. He wanted to test the struggle for existence by growing different species of protozoa in pairs, in the same container and with the same food supply. Gause published his findings in a book entitled *The Struggle for Existence* (1934),

26 THE GAME OF SPECIES

FIGURE 3.2 MacArthur's warblers. These five species of warbler prefer to forage in different parts of the tree, as Robert MacArthur demonstrated. In this way they segregate their niches, reducing competition between them. Illustration by artist Deborah Kaspari, showing the warblers grouped in a conifer like the 'Christmas tree' used by MacArthur to illustrate his findings. From the top and clockwise, the birds are males of Cape May Warbler *Setophaga tigrina*, Blackburnian Warbler *S. fusca*, Black-throated Green Warbler *S. virens*, Bay-breasted Warbler *S. castanea* and Yellow-rumped Warbler *S. coronata*. They are drawn roughly in their preferred area of the tree, with the exception of the Yellow-rumped Warbler, which frequents a lower level of foliage.

and in the last sentence of his Chapter V, which he devoted to the competition between these microscopic creatures, he summarised his conclusions:

> Owing to its advantages, mainly a greater value of the coefficient of multiplication, one of the species in a mixed population drives out the other entirely.

After Gause, many others carried out similar experiments, forcing different organisms, from yeasts to beetles, to compete, and the results led to a general rule: the species that competes best usually eliminates the other. Not always, but almost always. This was called *competitive exclusion*.

However, to some extent all these laboratory experiments are still forced, unnatural situations. What happens in nature? Are there known cases of competitive exclusion in natural communities? Yes, and Darwin reported what seems to be one of them, involving wild grasses:

> If turf which has long been mown, and the case would be the same with turf closely browsed by quadrupeds, be let to grow, the more vigorous plants gradually kill the less vigorous, though fully grown plants; thus out of twenty species grown on a little plot of mown turf (three feet by four) nine species perished, from the other species being allowed to grow up freely.

Of the many possible more recent examples, I have selected two marine stories to show that Gause was right not only about microbes.

First, let's go to the rocky and cloudy shores of the island of Great Cumbrae in the Firth of Clyde, Scotland. There, on the reefs washed by the Atlantic tides, lives a strange community of animals whose functioning the ecologist Joseph Connell wanted to understand in the middle of the twentieth century. He wondered how two species of barnacle, a type of crustacean which grows attached to rocks, could live together. The adults of *Chthamalus stellatus* grew on the rocks in a band above the area where the adults of the other barnacle, *Semibalanus balanoides* (known in Connell's day as *Balanus balanoides*), were found. Like a detective looking for clues, Connell noticed that in the lower area, the *Semibalanus* area, there were many tiny, very young *Chthamalus* ... but not a single adult. Was competition with the other species preventing them from reaching adulthood?

To find the answer, Connell devised an elegant series of experiments that confirmed his intuition: he discovered that *Semibalanus* was competing fiercely in the lower area, growing and spreading at full speed, even climbing over the young *Chthamalus*. Thus, *Semibalanus* was not allowing *Chthamalus*, which developed more slowly, to reach the adult stage. In this way, *Semibalanus* prevented *Chthamalus* from completing its life cycle, and so competitive exclusion was occurring.

With these observations, Connell showed that the niche of *Chthamalus* was being 'cut out' by *Semibalanus* on the lower coastal rocks. Thus, he proved that the niche of a species can be narrowed by competitive exclusion. The *Chthamalus* were only able to reach adulthood on that coast because their relentless competitors had a weakness: they could not withstand desiccation as well as *Chthamalus*. Thus, above the *Semibalanus* area, where the high tide

FIGURE 3.3 Rocky shores have taught us a lot about the game of species. Photographs of the Galician coasts of northwestern Spain: above, view of the Atlantic Ocean from Monteagudo, one of the Cíes Islands. Below, on the left, barnacles, almost all of the genus *Chthamalus*, but the largest and some others are *Austrominius modestus*, an invasive species from the seas of Australia and New Zealand. On the right, three limpets (*Patella*) and below them many snails (*Monodonta*) on a rock covered with reddish algae. Further down, on the left, Beadlet Anemones *Actinia equina* among *Monodonta* snails, and on the right, a very young European Green Crab *Carcinus maenas*, camouflaged among broken shells. Photographs by the author.

barely reached and conditions were drier, the *Chthamalus* managed to grow, free from having to compete with their enemies from below.

For our second example of competitive exclusion, we will go to the Pacific coast of the United States, to the rocks, also quite misty, of Makah (or Mukkaw) Bay, in the state of Washington. There, another ecologist, Robert Paine, was trying to understand the functioning of the coastal fauna. He discovered a curious food web that included a large predator, the fearsome Purple Sea Star *Pisaster ochraceus*. This voracious carnivore could eat any of the other animals in the community: chitons, limpets, mussels, barnacles (including goose barnacles) and even a predatory snail, *Thais*. All these species seemed to coexist in harmony – yet competition was unleashed and devastated the community when Paine wondered what would happen if he removed the starfish.

In June 1963, he chose an area of the shore measuring about 8 × 2 metres, and within a year of keeping it free of the sea stars, one barnacle species had grown to occupy almost all the available space on the rocks. By the second year, the barnacles were being covered by mussels, and it gradually became apparent that the mussels would eventually dominate. Most of the algae that covered the rock disappeared, because these animals left no room for them to grow. For the same reason, the chitons and limpets disappeared, also for lack of food (algae). In short, as soon as the starfish were removed, biodiversity began to decline through competitive exclusion. What had happened?

The key to understanding this is that the Purple Sea Stars ate, among other things, the most efficient competitors: barnacles and especially mussels. By eating them, the starfish made some of the space they occupied on the rock available to algae, chitons and limpets. In this way, the starfish kept the tougher competitors at bay, preventing them from spreading rapidly and eliminating the less competitive species. So this experiment shows us examples of competitive exclusion and the very important fact that predators can prevent this from happening – this is called keystone predation.

Because competitive exclusion has been observed both in the laboratory and in nature, and is also predicted by some mathematical models of interspecies competition, many ecologists in the mid-twentieth century took it so seriously that they promoted it to the 'principle of competitive exclusion'. Faced with this principle, some relaxed their judgement and tried to see competitive exclusion everywhere, even when there was no evidence for it. Without going to such an extreme, it is nonetheless clear that competitive exclusion can and does occur, both inside and outside the laboratory.

Waves, storms and other disasters

Sometimes dominant competitors are decimated not by their enemies, like the starfish, but by small natural catastrophes, generally referred to as disturbances.

These small-scale cataclysms can range from sudden weather changes that alter the temperature of a lake's water to floods, fires or storms that devastate certain sites. They act by resetting the history of the ecosystem, by damaging or eliminating the species that were winning the game. If disturbances occur very infrequently, a few dominant species will usually triumph in the community, and in their rise to ecological success at the site they are likely to exclude many others. But if disturbances are too frequent, only a handful of species that are highly resistant to such catastrophes will survive. So the greatest diversity will occur at an intermediate frequency of disturbance, because then we will have the dominant species, the resistant species and the others. This maintenance of maximum biodiversity at intermediate frequencies of disturbance is known as the *intermediate disturbance hypothesis*.

There are some very curious experiments that have confirmed this hypothesis. One of them was carried out at the end of the 1970s by the ecologist Wayne Sousa on a coastline in southern California full of pebbles. Several species of algae were growing on these stones, together with barnacles (funny how much we have learned about the game the game of species from barnacles!). But one disturbance agitated their lives: the swell of the winter storms, which was at times so strong that it turned the pebbles upside down. The smaller ones were turned over more often, as they weighed less, while the larger, heavier pebbles were seldom or never turned over. As predicted by the intermediate disturbance hypothesis, the pebbles with the highest number of species were those of intermediate size. The larger ones tended to be dominated by a few species, and the smaller ones were not very diverse either. Sousa glued some pebbles to the ground to make them more stable, and was thus able to confirm that the differences in diversity were due to the frequency of disturbance, in this case wave action.

It appears that intermediate disturbance is quite important for maintaining the biodiversity of some communities, such as coral reefs and tropical jungles, which are the most biodiverse marine and terrestrial ecosystems in the world. As Joseph Connell (the ecologist of the barnacle story) argued, strong tropical storms and hurricanes knock down jungle trees, or tear off chunks of rock from reefs, and in both cases open up clearings where the struggle for life begins again. He suggested that perhaps this was the mechanism that allowed so many species to coexist in both communities. Was he right?

In 1980, the shallow areas of the coral reefs in Discovery Bay, Jamaica, became more diverse after Hurricane Allen slowed the expansion of the most competitive species. And in 1982, the diversity of the Kona reefs (Hawaii) increased after a moderate storm. Other similar examples suggest that intermediate disturbance does indeed affect reefs. On the other hand, in the rainforests of Ghana, a 2009 study looked for evidence of the intermediate disturbance hypothesis in 2,504 one-hectare squares of rainforest containing 331,567 trees. The result? Evidence was found supporting the predictions of the hypothesis, but it was also found

FIGURE 3.4 Tropical rainforest in Borneo. Tropical rainforests are the most biodiverse terrestrial ecosystems, easily having more than 100 tree species in a single hectare. Part of this richness may be due to the fact that frequent storms cause trees to fall, opening up clearings where sunlight allows the growth of species that would not thrive in the shadows of these gigantic cathedrals of vegetation. Thanks to this process, more tree species can coexist. Photo by Dukeabruzzi.

that the effect of intermediate disturbance on biodiversity was very small in all but dry tropical forests.

Therefore, intermediate disturbance is of some importance at least in ecosystems that are subject to relatively frequent catastrophes (compared to the longevity of their inhabitants). The vegetation of the Mediterranean basin is an excellent example of this, as fires increase plant diversity and favour the growth of shrub communities that need a lot of sunlight. These communities are dominated by rockroses (*Cistus* species), Rosemary (*Salvia rosmarinus*) and thyme (*Thymus* species). These shrubs cannot grow in shade, so they practically disappear when a mature, closed forest develops, with trees such as the Holm Oak (*Quercus ilex/rotundifolia*) and Cork Oak (*Q. suber*).

***

Our journey through the world of ecological niches is coming to an end, and we have learned something very valuable, a basic rule of community organisation:

the species that make up a community occupy different niches. The differences may be obvious, or they may be subtle, but they are nearly always there. Possible exceptions to this principle, such as the 'paradox of the plankton' (see glossary), have turned out to be doubtful or non-existent on closer inspection.

Communities really do seem to assemble according to the roles played by the pieces, that is, how well the niches of incoming species fit together. For example, over the course of history, humans have brought many foreign birds to the islands of Hawaii, new species which have colonised the archipelago in a seemingly chaotic way. But in 1987, the ecologists Michael Moulton and Stuart Pimm revealed an underlying order. They found that 49 species of exotic birds were involved in a total of 111 introductions on five of the six islands, with 33 of these events ending in extinction. By analysing the body shapes of the birds, reduced to coordinates, they found that the species that had successfully invaded Hawaii's forests were more different from each other than would be expected if we had chosen them at random from all the arrivals. Thus, they found the smoking gun that proved that this community of birds had been assembled by each species occupying a distinct niche related to its body shape, a shape that makes it easier to manoeuvre in a particular ecological niche: tree branches, soil, etc.

The concept of the ecological niche will often accompany us over the next two chapters as we delve into the game of species, first on the simple board of the islands, and then on the vast board of the Red Queen. But before we embark on this adventure, we need to understand why I have divided the functioning of biodiversity into two boards.

Chapter 4

About the boards

In the mid-twentieth century, the engineer Frank Preston was busy carrying out research on the properties of glass in his Pennsylvania laboratory, but he was also an enthusiast of the game of species, and published his own ideas on the subject. Among other things, he wanted to know how the diversity of birds changes from a home garden to the whole planet, so he collected the data we see in Figure 4.1.

This plot is one of the best examples of one of the clearest rules of biodiversity: the species–area relationship. (Some think that this is the oldest ecological pattern known to humankind, but in fact another one was described earlier,

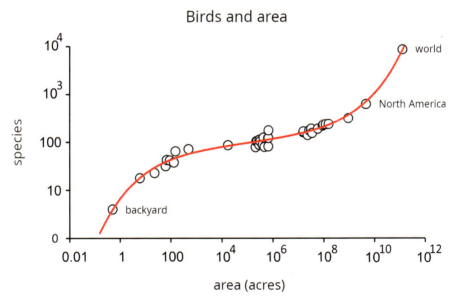

FIGURE 4.1 Bird species over increasing areas, from a backyard in North America to the world. The circles are the data and the red line is the trend. Species–area relationships such as this one are usually plotted on logarithmic axes, that is, from one unit to the next we do not add the same amount but multiply by the same number, in this case by 10 (vertical axis) and by 100 (horizontal axis).

the latitudinal gradient of diversity, which we will discuss in Chapter 7.) The species–area relationship means that the greater the area of nature we explore, the more species we usually find. This does not sound like a great discovery, quite the contrary, but the surprising fact is that the new species are added following a very particular curve, as shown in Figure 4.1. The number of species on the vertical axis may vary, but the shape of the curve remains almost identical whatever the type of organism. This universal pattern consists of three sections: a steep one at the beginning, a less steep one in the middle, and again a steep one at the end. This tripartite curve suggests that the game of species follows different rules as we increase the scale of the board on which we observe it.

We can think of this first section of the graph as corresponding to a local board on which the game is played, a habitat board. And we will understand what happens in this first section (and why we are probably not very interested in it) when we try, for example, to explore the diversity of insects found in a grassland habitat by sampling with a net. To do this, we will sweep equal areas of grass with the net and regularly examine the contents to record new findings. The number of species will increase rapidly as we sweep more and more of the grass, simply because we are catching more and more insects, and of as we increase the number of catches it is easy for a previously undetected species to appear. Nothing else is happening except that we are gradually discovering the diversity of insects in the grassland. In other words, the first part of the species–area curve corresponds mainly to the initial sampling of individuals needed to discover the diversity of a community.

Of course, in this first section, diversity rises more quickly or slowly depending on how abundant some species are compared to others. The rule of thumb that ecologists have learned from experience is that most species are *very rare* in their communities, and only a few manage to be abundant. We should not be too surprised by this, because the struggle for life in nature is so hard, as we saw in the first chapter, that only a handful of species are able to triumph in each place. In more mathematical terms, the local abundance of a species can be modelled as depending on a number of random factors with a multiplicative effect (because reproduction is a multiplicative process), and thus high numbers are very unlikely; therefore, few species are abundant. The arcane mysteries of the relative abundance of species have inspired hundreds of pages in the scientific literature, but they may not have helped us understand the game of species much more than this...

The final two sections of the species–area curve correspond to our islands board and to the Red Queen's board, respectively. In each section, diversity depends on different things. In the middle section, which is the universe of regions, provinces and most islands, diversity is controlled by processes acting on short timescales, say from days to millennia. By contrast, in the last section, the curve enters the vastness of what we will call Wallace's regions – huge areas,

ABOUT THE BOARDS 35

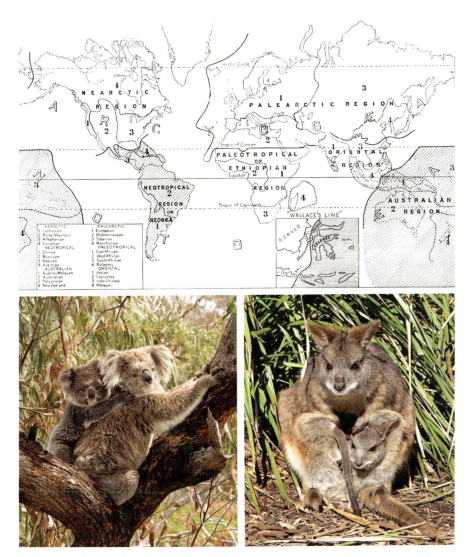

FIGURE 4.2 The faunal regions of the world, according to Sclater and Wallace, in a picture that is over 150 years old but barely distinguishable from the way that biogeographic realms are delineated today. In each of these vast areas, evolution has gone its own way, producing a multitude of characteristic forms – and the greater and older the isolation of the territory, the more distinct are its life forms. An example of this creation of alternative biological worlds is the Australian marsupial fauna, which includes the Koala *Phascolarctos cinereus* (left) and the Parma Wallaby *Notamacropus parma* (right). Isolated in Australia for tens of millions of years, without competition from placental mammals, marsupials evolved to occupy a variety of ecological niches. Photos by Ben Twist.

the size of large countries to continents or oceans, where biodiversity is born and dies over millions of years.

These enormous areas have produced most of their own species, as well as a wealth of unique genera, and even families and orders. The ornithologist Philip Sclater noticed this when he looked at the worldwide distribution of birds in the order Passeriformes. He found, for instance, that many families were endemic to the New World: Trochilidae, Tyrannidae, Parulidae... On the basis of these observations, in 1858 Sclater defined the major faunal provinces of the planet with such precision that they are still almost the same as those we use today. His work was extended by Alfred Russel Wallace, the co-discoverer of natural selection and a tenacious, globe-trotting naturalist, in a massive work entitled *The Geographical Distribution of Animals* (1876).

Each of Wallace's regions has its own more or less independent evolutionary history. Australia is a good example of this, as it has developed a very different fauna from the rest of the world during its approximately 40 million years of isolation from the other continents (Figure 4.2). Therefore, the evolution of species on the largest scale, that is, their origin and extinction, determines this last part of the species–area relationship. It is the Red Queen's board, as I call it for reasons I will reveal in the corresponding chapter.

The species–area curve unifies the boards of the game of species in space and also in time, as each board corresponds to a different spatial and temporal scale. A rule of thumb relates the boards to each other: the more species there are on one board, the more there tend to be on the other boards. For instance, if we were to look at a graph like that shown in Figure 4.1, but with data from South America, at local and regional scales we would have more bird species than in North America. Very diverse communities are normally found in very diverse regions, as if biodiversity were a music echoing across different scales of space and time. And indeed it is logical that a region with many species would have very diverse communities: more species can live in each place simply because there are more species in the region as a whole. This pattern, where local diversity tends to mirror regional diversity, implies that communities are not usually saturated with species, even in the richest regions. We will come back to the question of community saturation later. And now that we understand why the following chapters deal with two different boards, on different scales, let's see how the game of species is played on the first of them.

Chapter 5
The islands board

Now we will finally look at the game in action, not just at the individual pieces and their roles. We will start by looking at the less complex board. This board represents relatively small areas, over such short timescales that we can practically rule out the evolution of new species. Islands are the perfect setting for observing these simple games, as their isolation makes it easy to understand what is going on without the complications that would be added by the

FIGURE 5.1 Above, the islands of Hawaii (left), the archipelago furthest from the continents in the world, and the Canary Islands (right), both photographed by NASA. Below, a characteristic landscape of each archipelago: the island of Kauai and the Cañadas del Teide, respectively. The giant herbaceous plants in the foreground of the Cañadas are Mount Teide Buglosses *Echium wildpretii*, a species endemic to the island of Tenerife. Photographs by Lukas and Garavitofte, respectively.

presence of adjacent territories. The big idea running through this board is that the diversity of an island, or any similarly sized area, depends on the rate at which species arrive on it and the rate at which they go extinct on it. Over time, the species arriving on the island will tend to compensate for the species 'leaving' through extinction. The island's diversity will then remain more or less stable: it will have reached species equilibrium. Let's look at this in more detail.

Travellers who come and go

Imagine an underwater volcano that grows to emerge from the ocean and become an island. At first, the lava will form a rocky, desolate and barren terrain. After a while, by chance, the first settlers will arrive. They may be castaways washed ashore, or lost travellers flying in, or floating in the air like thistledown, or they may come on vegetation blown out to sea by storms. In any case, the more species that arrive from the surrounding area, the fewer will be left to arrive. Thus, as shown in Figure 5.2, the probability of a new species arriving (the immigration rate) decreases as more species arrive on the island.

But an island cannot support an infinite number of trees, nor infinite herbivores or carnivores. There is only space and resources for a certain number of individuals. So the more species there are, the less abundant each one can be. This decrease in abundance will make it easier for some species to go extinct simply because of bad luck (a bad year, a small catastrophe, etc.). In addition, as species accumulate, there is a greater chance that some will interfere with others, thus making the struggle for existence harder and increasing the risk of competitive exclusion. Thus, in Figure 5.2, the risk of extinction per species on the island increases with diversity.

As the graph shows, at some point immigration will equal extinction, and then the number of species will reach equilibrium. It is an equilibrium because it is maintained despite some oscillation in the number of species, as explained in the caption to Figure 5.2. At equilibrium, the game of species on the island will be a continuous balance of immigration and extinction, with species arriving and species going extinct. The composition of the community will change, but the number of species will remain close to the same value.

This brilliant idea was first proposed by the ecologist Eugene Munroe in the late 1940s, in two contributions that unfortunately went unnoticed by the scientific community, whose attention was focused on other issues and other personalities. It was not until the late 1960s that the concept of species equilibrium came to more general notice, when Robert MacArthur and Edward Wilson, of incipient fame and apparently unaware of Munroe's pioneering work, developed the same concept into one of the great milestones in understanding the game of species: the theory of island biogeography – biogeography is the

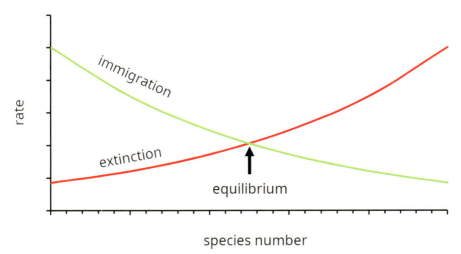

FIGURE 5.2 As the number of species on the island increases, the immigration rate (the probability that a species not already there will arrive) decreases and the extinction risk (the probability that one of the species will disappear from the island) increases, for the reasons explained in the text. The immigration and extinction curves intersect at a point which is the equilibrium point of the island's diversity. This equilibrium is stable: if the number of species falls below the equilibrium point, then diversity will increase because to the left of the equilibrium point immigration exceeds extinction; conversely, if diversity rises above the equilibrium point, then it will fall, because to the right of the equilibrium point extinction exceeds immigration. The result is that the number of species on the island will oscillate around the equilibrium number. This example shows curved lines as originally proposed by MacArthur and Wilson, but their concavity need not be exactly as shown here.

study of the geographical distribution of living things, including the distribution of their diversity.

Many were initially suspicious of MacArthur and Wilson's theory, and even harshly criticised it, but decades of evidence have shown that they and Munroe were right. In general, their idea that islands reach an equilibrium of diversity through a balance of immigration and extinction holds true, provided that no new species have evolved on the island.

Wilson himself and his student Daniel Simberloff carried out some unusual experiments that confirmed the equilibrium prediction. They chose six small islands in the Florida Keys, made up entirely of the tops of trees called mangroves, which were rooted directly in the sea mud and raised their branches above the waves. They first documented the diversity on these islands, each of which was home to between 20 and 50 species of insects and arachnids, out of the hundreds of arboreal invertebrate species in the region. Then they did something that is bound to provoke a reaction: they hired an exterminator to wipe out the fauna of the islands. The exterminator covered them with a kind of cloak and fumigated them thoroughly; later, according to eyewitnesses, he declared that he had never had so much fun in his life. The result was deserted islands, ready to be colonised again.

Wilson and Simberloff observed this colonisation over many months and found the expected trend: the number of species on each island first rose rapidly, then more slowly, and finally seemed to reach an equilibrium. Within 250 days, all islands except the most distant one had returned to about the same levels of diversity as before the spraying. After that time, the number of species changed incessantly, but always remained close to the equilibrium value. The islands had reached a changing, dynamic equilibrium of biodiversity, just as the theory predicted, as some species arrived and others went extinct, at a rate of one species every one or two days on the islands located about 200 metres offshore. Many of the extinctions were of recent colonisers, but species that had been on each island for longer also disappeared frequently.

Curiously, in many cases the species were different from those that had been found on the islands before the spraying; for example, the ants reached more or less the same number of species as before, but some species were different, as if each species could easily be replaced by another. Moreover, some islands had slightly more species than before, which was to be expected because it takes time for species to go extinct, for bad luck to occur and for vulnerable species to decline, and so the return to equilibrium was somewhat delayed.

There are many other cases we could review. For instance, the island of Krakatoa, in the Sunda Strait between Sumatra and Java, exploded in a volcanic eruption in August 1883. Half of the island was destroyed and the other half was covered in a fiery layer of volcanic ash and pumice some 30–60 metres

FIGURE 5.3 The island of Surtsey, in southern Iceland, emerged from the ocean amid volcanic eruptions on 14 November 1963. Sixteen days later, NOAA took the photograph on the left. Eruptions continued until 5 June 1967, when the first plants, Arctic Sea Rocket *Cakile arctica*, began to grow on the new land. Since then, scientists have watched how life has colonised this islet. The Atlantic Puffin *Fratercula arctica* (right) nested there for the first time in 2004. Photograph by Richard Bartz.

thick. After this catastrophe, life quickly took hold again, and by 1908 there were already 13 species of terrestrial and freshwater birds living on Krakatoa. Between 1919 and 1921, the number of bird species increased to 31. And this level seems to be the equilibrium, because more than a decade later, in 1932–1934, there were 30 species, virtually the same number. MacArthur and Wilson used this example to calculate the decrease in immigration rate and the increase in extinction rate predicted by their theory, which was supported by the results they obtained.

The beauty of the theory of island biogeography is not just its simplicity, but that it can help us understand many different phenomena. For example, why do islands typically have reduced biodiversity? Even in Darwin's time, it was clear that islands tend to have fewer species than an area of the same size on the nearest mainland. And the further away the island, the greater its impoverishment of species. This 'distance effect' is evident in the islands of Oceania and the Pacific, many of which have received species mainly from New Guinea. MacArthur and Wilson divided these islands into three groups: near (less than 500 miles from New Guinea), intermediate (between 500 and 2,000 miles) and distant (more than 2,000 miles). They found that, given the same area, there were usually more species of land and freshwater birds on the near islands than on the intermediate ones, and more on the intermediate than on the distant islands. For instance, the islands and archipelagos of Seram (Ceram), Fiji and Hawaii are all similar in area, between 6,500 and 10,000 square miles (17,000–26,000 square kilometres), and because of their distance from New Guinea, they are classified as near, intermediate and distant, respectively. And it is in this order that their number of bird species decreases. The same is true of the Kai (Kei) Islands, Rennell Island and the Society Islands. Why does this happen?

Very easy. If an island is far away, it will be more difficult for organisms from the mainland to get there. Therefore, the immigration rate will be lower, that is, fewer new species will arrive per unit time than on a island close to the mainland. In addition, distance also makes it more difficult for individuals to arrive that can save an island species from extinction (this is the 'rescue effect'), which means that we would expect a higher extinction rate on distant islands than on nearby islands. So on a remote island we will have less immigration, more extinction, and, as a result, a lower value of the diversity equilibrium than if the island were close to the mainland. This is an elegant explanation of the distance effect.

It is therefore understandable that even an island close to the mainland is impoverished in species compared to a mainland area of the same size. Imagine a continental area of the same size as an island. As there is no sea around it, species will very easily reach the interior of that area. This means a huge immigration rate and a minimal risk of extinction, because the rescue effect will be very strong – there may even be species that are often there, coming from outside, without breeding in that area. As a result, there will be more species on that patch of mainland than on the island, even if it is close to the coast. This is the reason why island biodiversity is impoverished! How could it not be, since species arrive with difficulty but easily become extinct?

Another surprise of island biogeography was its explanation of why large islands generally have more species. A major ingredient in understanding this is the variety of habitats, which tends to be greater the larger the island. An islet may have only low-lying rocks, but a large island may have snow-capped mountains, waterfalls, rainforest and more. This means that more immigrant species can find a place to live on larger islands. For example, as the range of island habitats increases, so does the diversity of non-marine birds in the Canary Islands, birds in the Aegean islands, snails on certain islands in the Bahamas, and plants in the Galápagos archipelago. In addition, a large island will be able to intercept more travellers than a small one, simply because of its size, and this will result in more immigration and less extinction (owing to the rescue effect). Taken together, this will give us an equilibrium with more diversity on larger islands. The shape of the island can modify these considerations to some extent, for instance by having more or less coastline on the side from which more travellers arrive. These considerations also apply to a continental area and help us to understand why there are more species in larger areas in the middle part of the species–area curve (see Figure 4.1 again).

A free buffet

As proposed by MacArthur and Wilson, the theory of island biogeography does not require changes in species niches. For the risk of extinction to increase with diversity, it is sufficient for the island to be able to support only a certain number of individuals; then, if there are many species, there will necessarily be fewer individuals of each species, thus facilitating extinction. For the risk of extinction to increase per species as diversity increases, species niches do not have to change. But they do change, and in ways that also facilitate equilibrium.

The niches of island organisms become wider the fewer similar species there are on their island. And vice versa, as diversity accumulates, niches become narrower. This is the principle of ecological release: when a species is freed from its competitors, it expands its niche. It looks for food in places it previously avoided, or grows in habitats where another species prevented it from doing so, and so on.

Ecological release teaches us that the niche of a species is often limited, narrowed, by its competitors – each such 'narrowing' would be an example of competitive exclusion. This is why many continental species expand their niches when they find themselves in the species-poor environment of an island. For instance, the Rock Sparrow *Petronia petronia* lives on rocky cliffs in Morocco and Spain, but in the Canary Islands it also occupied villages. Until the arrival of

FIGURE 5.4 Birds and anole lizards of the Caribbean archipelago tend to expand their ecological niches when there are few competitors on the island. From left to right, and from top to bottom: two Bananaquits *Coereba flaveola*; a Jamaican Tody *Todus todus* (endemic to Jamaica); the Cuban Green Anole *Anolis porcatus*; and the Tocororo *Priotelus temnurus*, the national bird of Cuba, where it is endemic. Photographs by Leon-bojarczuk, Charlesjsharp, Thomas H. Brown and Laura Gooch, respectively.

the Spanish Sparrow *Passer hispaniolensis*, which has been spreading across the islands since the end of the nineteenth century, displacing the Rock Sparrow from the villages.

There is a veritable arsenal of observations supporting ecological release on islands, and we will now visit the West Indies to discover two wonderful examples. The first concerns the land birds of this archipelago. The ornithologists George Cox and Robert Ricklefs systematically observed the diversity of birds in six habitat classes on the islands, counting the number of species seen or heard in a fixed period of time. With their data in hand, they asked how species niches changed as the number of species on the island increased. The results were very clear: the fewer species on the island, the more habitats each species tended to occupy. In other words, freed from competitors, these Caribbean birds expanded their niches. As an island had more and more bird species, the average range of habitats in which each species could be found decreased. Their competitors were cutting out portions of their niches, forcing each species to focus on a few habitats where their skills were best suited to survive. But if one of these bird species is cornered by its competitors in a single habitat, and that habitat becomes scarce, that unlucky species may become so scarce that any stroke of bad luck could mean extinction. So the narrowing of niches increases the risk of extinction on the island.

Our second Caribbean story of ecological release was documented in detail by the ecologist Bradford Lister with anole lizards (genus *Anolis*). On islands with less diversity of these lizards, each species expanded its niche in a variety of ways: colonising new habitats, climbing on branches of different heights to rest, living in different microclimates, or hunting prey of a wider range of sizes. Anoles were taking advantage of a buffet of untapped possibilities. This relaxed lifestyle gradually ended as their island became home to more and more anole species, since competitors occupied different corners of the buffet.

The pinnacle of this 'anything goes' of ecological release on islands may well be the Gough Finch *Rowettia goughensis*. The ornithologist David Lack dubbed it 'the ultimate, all-purpose bird'. It lives only on Gough Island in the South Atlantic, far removed from the rest of the world, being some 3,000 kilometres from both South America and Africa. The island rises up to 910 metres above sea level and has a good variety of habitats, from coastal cliffs to mountain vegetation, and from peat bogs to bracken and heathland. And he only passerine that occurs there regularly is the Gough Finch, which uses every available habitat. It even forages on the shore at low tide, and eats seeds, fruit, insects, carrion and more; indeed, it has made the whole island its buffet.

The filling of niches on an island may also slow down the establishment of new species. If there are already many species on the island, there are likely to be many occupied niches, so a new species will find it harder to find a relatively unexploited niche where it can survive without facing stiff competition. This means that the more species there are on an island, the more likely it is that immigrant species will fail to colonise it. In other words, diversity protects against invasion. This idea was popularised by the ecologist Charles Elton, in an assessment of whether an ecosystem would be more or less susceptible to invasion by alien species.

Biological resistance to invasion is usually observed on a small scale, say for the 'neighbourhood' of plants and animals. This has been confirmed in a large number of careful experiments, in grassy fields, rocky shores and other habitats. On a larger scale, however, the surprise is that the relationship seems to be reversed, with more invasive species found in regions of higher diversity. This happens despite the fact that experiments show that diversity confers resistance to invasion at small scales in these same regions. How can we explain this paradox? I have to leave the question open and simply point out that there are many reasonable solutions that have nothing to do with the direct effect of biodiversity: more attempts at introduction in these areas, a wider range of habitats, more environment modification… Anyway, let's take the rule demonstrated by the experiments: at local scale, biodiversity tends to block colonisation, and so it could promote the mechanism that leads to equilibrium.

***

From this journey through the world of islands, we bring back a powerful aid to deciphering the game of species: the theory of island biogeography, or in other words, the balance between immigration and extinction. Although it applies to islands, it can be extrapolated to continental areas and even to the aquatic environment; only the rate of arrival and 'departure' of species would change.

The main objection to this theory is that it does not take into account the origin of new species. But didn't we say at the beginning of this chapter that it doesn't, by assumption? Nevertheless, it is very common for species to originate on islands, especially if they are large and have many habitats. MacArthur and Wilson knew this, but they didn't want to complicate their model any further. To include speciation, the ecologist Robert Whittaker and some of his collaborators proposed what they called 'general dynamic theory'. This even considers the geological history of islands, which often erode over a few million years, losing the variety of habitats that helped maintain biodiversity. In this model, the number of species on an island depends strongly on its area and age, in a formula that predicts quite well the percentage of endemic species on many islands.

In this way, through the islands we can glimpse the game of species in all its complexity. How does the game change when we include species that are born and die, but in regions as large as our planet can contain? Come with me, in the next chapter, into the vastness of oceans and continents, into the abyss of evolutionary time. We will explore what rules life has followed to unfold the kaleidoscope of its diversity over the eons in every corner of the biosphere.

Chapter 6

The Red Queen's board

We contemplate on this board the amazing story of evolution, the odyssey of strange primitive creatures being born and dying out, of species succeeding each other as the geological eras go by, until they gradually create the living world we experience today. This is called macroevolution, evolution on the largest possible scale, in both space and time.

Many naturalists still believe that macroevolution works the way Darwin thought it did: new species are formed gradually and survive only at the cost of replacing the old ones, because they are more efficient and perfected: 'Thus the appearance of new forms and the disappearance of old forms ... are bound together.' This implies a balance between new species arising and old species becoming extinct. The result would be a diversity equilibrium similar to that of islands but on a vast scale, for continents and oceans, in a bird's-eye view over millions of years. However, in the next sentence of *On the Origin of Species* Darwin tells us something that seems to contradict the previous idea:

> In certain flourishing groups, the number of new specific forms which have been produced within a given time is probably greater than that of the old forms which have been exterminated.

This apparent contradiction serves to present the two alternatives offered by the game of species on this board: is life a balance of species, or is it a relentless accumulation of species? Is the planet's biodiversity constantly increasing, or is it in equilibrium?

The balance of evolution

In the realm of millions of years, why should we expect diversity to reach an equilibrium, with a balance between the birth of species and their extinction. We have good reasons to think that this is the case, processes working so that, as diversity accumulates, speciation becomes more difficult and extinction easier, bringing the two closer together until they equalise and an equilibrium is reached (Figure 6.1).

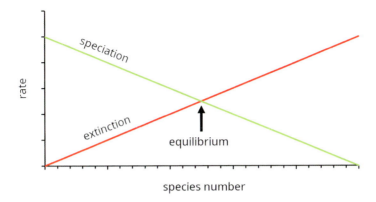

FIGURE 6.1 Above, the evolutionary balance of biodiversity. As the number of species increases, speciation (understood as the probability that one species will give rise to another) decreases. Conversely, extinction (the risk of a species going extinct) increases. When speciation and extinction are balanced, diversity reaches a point of equilibrium. This equilibrium is stable: if the number of species falls below it, it will increase (because to the left of the equilibrium point, speciation is greater than extinction), and if it rises above it, it will decrease (because to the right, extinction exceeds speciation). The lines do not have to be straight, as in this example.

Below, the Mundo River Butterwort *Pinguicula mundi*, a carnivorous plant, has trapped a fly with the sticky glands that cover its leaves. Because of its strict habitat requirements, this species is a good example of why species with very narrow niches face a high extinction risk. It lives only on shady, damp limestone walls, a habitat so specific that sudden climate change could eliminate it, driving this small plant to extinction. It is an Iberian endemic, with only a few populations in the Serranía de Cuenca, Sierra de Alcaraz and Sierra del Calar del Mundo. Photograph by the author.

First, the more species there are, the harder it is for evolution to add a new species. This is because with many species there will be many occupied niches. So it will be more difficult for a new species to find an ecological niche that is relatively free of competition. Therefore, the greater the diversity, the less likely it is that one species will give rise to another (this is analogous to the invasion resistance described in Chapter 5, but on an evolutionary scale).

Second, as species accumulate, the same thing will happen as on an island: the risk of extinction for each species will increase, on average. The reasons for this are the same as those on that board: if the region can only support a certain number of individuals, then the more species there are, the less abundant each one will be. This decline in abundance will make it easier for some species to go extinct, by accident or bad luck. Furthermore, as we saw in Chapter 5, the more species that coexist, the narrower their niches, and this will also make survival more difficult (Figure 6.1).

Let's see if all these expectations are met in nature. There is very convincing evidence that speciation tends to be facilitated when there are few species and slowed down when there are many, as shown by the green line in Figure 6.1. This is true for islands, which, as we know from the previous chapter, have few species compared to the mainland. This means a lack of competitors, so the islanders will expand their niches (see *A free buffet*, Chapter 5). Thus, on an island, animals and plants can try out new ways of living that would never be worthwhile on the mainland, where they would face stiff competition. And because of the geographical isolation typical of islands, it will be easy for natural selection to transform the organisms that occupy new niches into new species. Hence, remote islands are breeding grounds for species, not only because of their isolation (see *Geographic isolation*, Chapter 2), but also because their low diversity facilitates niche shifts for their inhabitants, thus promoting speciation by natural selection.

The freedom to explore new niches has produced on the islands an unparalleled variety of strange creatures, organisms in whose appearance and way of life we can sense the full force and beauty of evolution by natural selection. On the continents, the finches (family Fringillidae) have not produced any species that feeds like a woodpecker, and yet in Hawaii there is a group of finches known as honeycreepers that includes the 'Akiapola'au *Hemignathus wilsoni*, which removes the bark from branches with its strange beak in search of insects to eat, just as woodpeckers do. On the mainland, finches have produced nothing like a hummingbird, while on these islands they have evolved into the 'I'iwi *Drepanis coccinea*, one of the archipelago's nectar-sipping birds.

The repertoire of biological oddities found on islands paints a picture of a world where evolution has broken free from the shackles of much fiercer competition on the continents: there are carnivorous caterpillars of geometrid butterflies that hunt insects with their legs (*Eupithecia palikea*, Hawaii); 'vampire' finches which

FIGURE 6.2 Sampler of island rarities. Above, a male Marine Iguana *Amblyrhynchus cristatus* subsp. *venustissimus* with breeding colouration, on Española, Galápagos. Below left, a Kea *Nestor notabilis* during a snowfall in the mountains of New Zealand. Next to it, an 'I'iwi *Drepanis coccinea* from Hawaii. Below that, a Kākāpō *Strigops habroptilus* from New Zealand. Photographs by Ben Twist, Alan Liefting, US Geological Survey and Chris Birmingham, respectively.

peck at the wings of seabirds and lick the resulting blood (the Sharp-beaked Ground-Finch *Geospiza difficilis*, Galápagos); flower-pollinating lizards (Lilford's Wall Lizard *Podarcis lilfordi*, found on islets off Menorca); cactus-eating iguanas that occupy the niche of herbivorous mammals (all three species of the genus *Conolophus*, Galápagos); other iguanas that swim and eat algae (the Galápagos Marine Iguana *Amblyrhynchus cristatus*); bats that swoop down to hunt giant crickets (New Zealand, perhaps the most unusual land of all); alpine parrots (the Kea *Nestor notabilis*, New Zealand); large, flightless, nocturnal, whiskered parrots that burrow and forage in the undergrowth (the Kākāpō *Strigops habroptilus*, New Zealand); penguins that breed in dense forests (the Tawaki or Fiordland Penguin *Eudyptes pachyrhynchus*, New Zealand); a lion-sized carnivorous lizard (the Komodo Dragon *Varanus komodoensis*, from some Indonesian islands); large tree lizards that climb with a prehensile tail, eat leaves and form monkey-like social groups (the Solomon Islands Skink *Corucia zebrata*).

Islands aside, other situations of low diversity and high probability of speciation are the scenarios that remain after the great annihilations of the history of life, the mass extinctions. There have been about five or six such catastrophes in the last 600 million years, each with its own cause or causes: meteorite impact, massive volcanic eruptions, sudden glaciation... In each of these apocalypses on the evolutionary board, at least half of the world's species became extinct in a short geological time, on the order of a million years and usually much less. The devastation of competitors after a mass extinction means a multitude of unoccupied niches and thus many speciation opportunities for the survivors. A few million years after each catastrophe the rate of origin of new species accelerates, and many new genera and families emerge. This 'speciation rebound', called evolutionary radiation, slows down as species accumulate. Of the known examples of this phenomenon, perhaps the most striking is the diversification of mammals that followed the extinction of the great dinosaurs.

Mammals have been around for about the same time as dinosaurs, but for more than 100 million years they were mainly small and apparently unspecialised animals. Throughout that time, it seems that all niches for land animals over 15 kilograms were occupied by dinosaurs and other reptiles. This reptilian dominance ended 66 million years ago with the most famous of the mass extinctions, the Cretaceous–Palaeogene (K–Pg) extinction event. Fossils show that, in the millions of years that followed, there was an evolutionary 'explosion' of mammals. Speciation accelerated and, without the colossal dinosaurs to compete with, the humble mammals were finally able to evolve to great sizes. And they soon did.

The largest mammal known before the extinction, as I write these lines, was *Repenomamus*, which weighed a maximum of about 14 kilograms. But only about two million years after the K–Pg extinction there was *Pantolambda*, the size of a sheep. And some three million years later came a relative known as

Barylambda, which at about 650 kilograms exemplifies the drift of mammals towards bulky herbivore niches. Although some scientists have questioned whether the extinction of the dinosaurs accelerated the evolution of mammals, new analyses based on molecular data confirm the idea, and the fossils leave no doubt.

Thus, speciation is greatly favoured when there are few species, either on islands or after mass extinctions. What about extinction, the red line in Figure 6.1? Well, it doesn't even need to be a rising line: as long as it crosses the line of speciation, we have an evolutionary equilibrium of biodiversity. Extinction could perfectly well be a flat line, that is, independent of diversity. Let's stress this: for species to be in equilibrium on an evolutionary scale, extinction does not have to depend on biodiversity.

There is some evidence that diversity does in fact increase the rate of species extinction, but there seems to have been little interest in exploring this issue. The example I know best comes from one of the most extraordinary fossil sites on the planet: the Rambla de Valdemiedes in Murero (Zaragoza, Spain). Since its discovery by the palaeontologist Édouard de Verneuil in 1862, this site has yielded tens of thousands of fossils of trilobites, cute marine invertebrates that are now extinct, and whose remains in Murero are found in strata deposited between 511 and 503 million years ago, during the Cambrian period. Some years ago I published a research paper describing a new method that 'shuffles' the observed distribution of fossils, using hundreds of computer simulations, to detect signals of biodiversity regulation. I applied the method to data from the Rambla de Valdemiedes and found that its trilobite species are not randomly distributed along the sequence of strata, but follow this rule: the more species there were, the higher the extinction risk of each one. I also found indications that diversity slowed down speciation, thus providing the two trends of Figure 6.1.

With exceptions such as that found at Murero, looking at fossil species is often not very practical for figuring out how they evolved, not least because poor preservation means that it is often difficult to distinguish which species a fossil belongs to. A compromise solution is to focus on genera, a level of classification very close to species and much easier to detect and identify in rocks. At the genus level, the lines in Figure 6.1 hold true for marine invertebrates, as tested by the palaeontologist Michael Foote on the basis of tens of thousands of finds of their fossils (which are the most common of all), almost from their origin over nearly 500 million years. Marine invertebrates, especially snails, bivalves and brachiopods, also narrowed their niches at times when there were more genera, which is consistent with evolutionary equilibrium. We cannot assume that these results at the genus level tell us what is happening at the species level (although this has been confirmed for molluscs), but they still provide good indirect evidence in favour of evolutionary equilibrium.

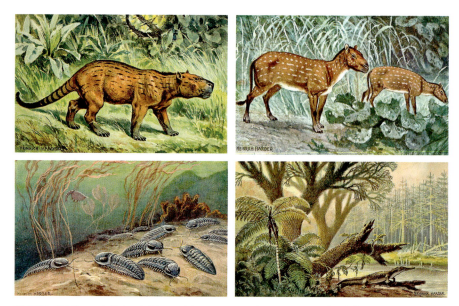

FIGURE 6.3 From left to right, and from top to bottom: *Pantolambda* (a mammal that evolved shortly after the extinction of the great dinosaurs); *Eohippus* (one of the earliest members of the horse family, which shows the patterns found by Quental and Marshall – see text); trilobites (marine arthropods very common in the early Palaeozoic era); and a landscape with the amphibian *Archegosaurus* in a forest full of primitive seedless plants, illustrating the colonisation of the land by animals. Illustrations for stickers by the palaeo-artist Heinrich Harder, ca. 1920.

Equilibrium lines also occur in the terrestrial mammal genera that lived after the dinosaur era, as evolutionary biologists Tiago Quental and Charles Marshall found by analysing 19 mammal families. They also found something very strange: although the conditions for equilibrium were apparently met, these families almost never reached it. The reason is subtle, and will reveal the ruler of this board of the game of species.

Quental and Marshall observed that, as a family aged, it usually developed 'ailments' of old age, which could be summarised as 'deteriorating health' (its genera became extinct more easily) and 'loss of fertility' (it was harder for new genera to emerge). The effect of these ailments was to increase extinction rate and decrease speciation rate, so that the theoretical equilibrium point was gradually reduced. What is the reason for this evolutionary 'senility' of families? The answer seems to lie in the fact that the environment becomes increasingly hostile to species as they age. This is the logical consequence of evolution's constant refinement of the species' enemies (competitors, predators and parasites); moreover, climate and other conditions can also change and cause harm.

All these problems trigger a race in which species must evolve not just to adapt better, but to survive. It is like Chapter 2 of Lewis Carroll's *Through the Looking-Glass*, in which the Red Queen explains to Alice that 'here, you see, it takes all the running you can do, to keep in the same place.' These words inspired the biologist Leigh Van Valen to give a name to this evolutionary dynamic: the *Red Queen hypothesis*. It seems to be a consequence of the struggle for survival but extended to the entire duration of the species. Seen in this light, Darwin was already thinking in terms of the Red Queen, which is why he assumed that the new species would be more efficient than the old ones, better able to cope with increasingly deadly predators, with more tenacious parasites, and so on. One sign of the existence of this never-ending race is that, since the beginning of the evolution of marine invertebrates, the genera that have emerged over the geological eras have been, on average, increasingly durable, that is, more resistant to extinction.

For all the above reasons, it seems reasonable to expect a general trend towards an equilibrium in species diversity on a scale of millions of years. Spectacular support for this expectation comes from the cliffs of Mistaken Point in Newfoundland, Canada. The strata that rise up there against the crash of the waves bear witness to ancient volcanic eruptions that buried the fossilised imprints of strange organisms under layers of ash. These beautifully preserved remains are the Avalon biota, the first relatively complex multicellular beings known to have evolved on the planet. They were an extravagant collection of creatures that grew in the darkness of the deep sea, some 565 million years ago, in the Ediacaran period (Precambrian). Forms like a fern leaf abounded, immobile, attached to the seabed, with branching arranged in a fractal pattern (the same type of branching formed at different scales of size). No animal known today builds its body in this way. Because of this bizarre anatomy, these fractal organisms, called rangeomorphs, may not have been animals at all, but a separate kingdom of creatures, now extinct, that the palaeontologist Adolf Seilacher called 'vendobionts'. They must have fed on nutrients or food particles in the water, as they show no signs of having had a mouth, and the abysses in which they lived had no light for photosynthesis.

Amazingly, the diversity of this first experiment in complex life on Earth was already typical of its modern ecological counterparts, namely the sea-floor fauna of the continental slopes. The Avalon communities yield diversity values within the range of these modern analogues, rather towards the less diverse end. Finding this enormous conservatism in biodiversity, in organisms that are perhaps the strangest life forms that evolution has created on this planet, suggests that the rules of the game of species are truly universal, that they do not change over hundreds of millions of years, even in organisms that seem completely alien to those we are familiar with today.

FIGURE 6.4 Fossils of unusual marine life are drawn in relief on this surface of a stratum at Mistaken Point, Newfoundland. They are the remains of the oldest known communities of relatively complex multicellular life. Despite their very remote age and the rarity of their forms, the diversity of these communities was within the typical range of their modern ecological counterparts. Pictured here are many comb-shaped *Fractofusus* and some *Charniodiscus*, which lived anchored to the sea floor by a disc with a leaf-shaped blade. Photograph courtesy of Alex Liu, University of Cambridge.

Ways of living

The equilibrium of species diversity is only half of the game as it is played on the Red Queen's board. If you are familiar with the history of life, you will know that over a geological timescale there have been a few episodes where the global number of species has dramatically increased in a short time. There have not been many such episodes, but there they are, proving that the game on this board also involves the expansion of life into new environments and new ways of surviving. A speciation boom, or evolutionary radiation, often accompanies the entry into these still untouched dimensions of ecological space – as expected, given the lack of competition. This happens after a mass extinction, just as on a newly formed remote island.

One of these dramatic increases in biodiversity was the colonisation of land. Plants began to tentatively spread along the shores of bodies of water at least 470 million years ago, as evidenced by fossil spores resembling those of modern liverworts, extremely primitive plants. Spores attributable to the more complex vascular plants date from around 450 million years ago. A hundred million years later, there were gigantic forests full of huge seedless plants in the clubmoss and horsetail groups, both of which are now marginal in the world's flora. Since then, little by little, plants have gradually colonised increasingly hostile lands (very dry, cold, torrid…), expanding their domains and now numbering some 400,000 species.

Animals settled on land at about the same time as the plants on which they depend, but they became much more diverse, with many more species. This great diversity has arisen especially thanks to insects, which alone are estimated to number more than a million species. Without the adaptations that allowed plants and animals to live on land, the diversity of life would certainly be much lower.

Life has also expanded its range of habitats in more subtle ways. The palaeontologist Richard Bambach and his colleagues found this out by studying how the niches of the shallow-sea fauna have changed over nearly 600 million years. They classified the possible niches that a marine animal could occupy, based on its position relative to the sea floor as well as its mobility and feeding strategy. They used this classification to assign niches to marine animals in the fossil record, and then analysed how the number of niches had changed over time. Let's look at what they found in the fauna that provides the most reliable information, the animals that fossilise most easily because they have hard parts such as shells. In this fauna they found 19 niches from the early Cambrian period (some 520 million years ago), 30 from the Ordovician (487–443 million years ago), and 62 in the recent fauna whose hard parts are regularly preserved in sediments. There is therefore no doubt that marine animals with skeletons have expanded their range of niches.

As a result of this evolution of new niches, the number of animal species typically found in shallow marine communities has increased significantly throughout the history of life. Bambach found that this increase in diversity was primarily achieved not by 'packing' more species into the same modes of life, but by adding species with new ways of living.

***

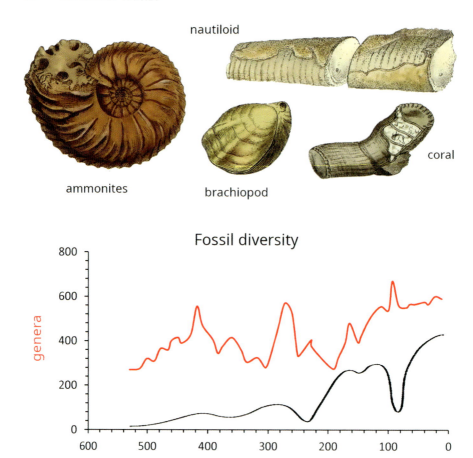

FIGURE 6.5 Above, some marine invertebrate fossils, illustrated by James Sowerby (1812). Below, curves of fossil diversity spanning almost from the origin of the animals. The red line represents the global diversity of marine invertebrate genera, according to a 2008 paper. The black line is the earliest known fossil diversity curve, published by John Phillips in 1860 and here scanned and digitally stretched for comparison with the more modern one, although it does not follow the vertical scale because it refers to species rather than genera. Phillips used the fossils of marine animals known from the British Isles to construct this curve, from which he divided the history of life into three eras: Palaeozoic, Mesozoic and Cenozoic, separated by two major biodiversity crises. The diversity of fossils we know from each period depends critically on the amount of sedimentary rock preserved at the time (the more rock, the more species we can find). This bias has been taken into account in both curves, and although the red one was obtained using a much more complex method to correct for it than the black one, and they are almost 150 years apart, the curious thing is that, if we ignore the details, the two curves are similar in showing a net increase in diversity, slow, irregular and seemingly accelerating from the last two hundred million years.

To recapitulate, the Red Queen's board gives us reason to expect the two things that Darwin brought together in apparent contradiction in that paragraph from *On the Origin of Species* with which we began this chapter. On the one hand, there is a tendency towards an equilibrium of diversity; on the other, diversity tends to increase. We could call this final level of the game of species the 'expansive equilibrium': the number of species tends to balance within the modes of life already occupied, but it also increases as species with new ways of living appear.

According to this expansive equilibrium, we would expect diversity to grow gradually, with no major changes in pace, but the red line in Figure 6.5 paints a different and rather chaotic picture. It shows that marine invertebrate diversity has experienced significant ups and downs, and long periods of decline, throughout virtually all of its evolutionary history. These features do not seem to fit with the concept of expansive equilibrium. What could this mean? Probably that the game of species on this board is often disturbed by planetary events that affect biodiversity: climatic changes, sea-level rise and fall, asteroid impacts... Some of these events create new habitable space, where new species can evolve, while others bring extinctions of varying magnitude, or unfavourable epochs lasting millions of years. The graph thus suggests that diversity not only obeys its own rules, but is swayed by the vicissitudes of the planet. In 2001, the palaeontologist Anthony Barnosky dubbed this notion the *Court Jester hypothesis*, because of the whimsical leaps and bounds typical of such a character.

Despite these environmental ups and downs, the diversity curve in Figure 6.5 is not as far from our expectations as it might seem at first glance. Diversity has remained stable, within the same order of magnitude, and there is a slight upward trend, especially over the last couple of hundred million years. This is broadly consistent with the idea that diversity tends to equilibrium and to grow gradually. But the unpredictable jumps of the Court Jester, that is, the more or less random changes in the Earth system, disrupt this functioning. In ancient Greece, Democritus said that everything that exists is the result of chance and necessity. These two factors come together on the Red Queen's board, to make the history of life as inexorable as it is capricious.

Chapter 7
Redux

In the following pages, our exploration of the game of species will come to an end. Many complexities remain to be explored in greater depth, but for the moment it is wise to stop here, for too often uncertainty and guesswork would accompany us if we delved deeper into the rules of the game. However, we have not yet attempted to answer what are perhaps the two simplest questions one can ask about this game. I posed them in the first paragraph of the preface, but I have reserved addressing them for this penultimate chapter: why do some places have more diversity than others, and why do some groups of organisms have more species than others?

The enigma of the tropics

We will try to answer the first question by looking into one of the classic questions of biodiversity: why do the tropics harbour so many species?

> Thus, the nearer we approach the tropics, the greater is the increase in the variety of structure, grace of form, and mixture of colours, and also the perpetual youth and vigour of organic life.

Since Alexander von Humboldt published these words in 1807, the increase in biodiversity towards the tropics, that is, the latitudinal gradient of biodiversity, has become one of the most solid rules of ecology. We can be almost certain to find many more species as we move from the poles to the equator, regionally and also in local communities, as long as we look at a sufficiently large group – for instance, penguins do not follow this rule, as they are more diverse towards the poles, but birds as a whole do, as do groups such as mammals, animals, plants… This is not a recent quirk of modern life, as the latitudinal gradient has existed at least during several epochs in Earth's history. Why is this pattern here?

In fact, it is hardly surprising that the tropics have the highest biodiversity, because there are clearly factors in the tropics that favour speciation, hinder extinction, or both, ultimately promoting the regional accumulation of more

FIGURE 7.1 *The River of Light*, a landscape inspired by the tropics of South America, by Frederic Edwin Church (1887).

species than in non-tropical regions. Indeed, all three options occur among the mammals: some groups are more diverse in the tropics because they have experienced higher rates of speciation there (diprotodontid marsupials, artiodactyls and soricomorphs), others because they have had lower rates of extinction (primates and lagomorphs), and others for both reasons (bats and rodents). Additional examples of a higher rate of tropical speciation, or a lower risk of tropical extinction, or both, have been found in birds, invertebrates, plants, and other groups. For this reason, using a popular metaphor devised by the botanist George Stebbins, it can be said that the tropics are a 'cradle' and a 'museum' of species.

Cradle? Yes, because tropical regions contain a greater variety of habitats: jungles, savannahs, snow-capped mountain, coral reefs... The wide range of environments and living conditions provides opportunities for ecological speciation. In support of this idea, we note that most of the bird and plant species of the South American tropics originated by adapting to different mountain environments created by the rise of the Andes, that huge mountain range. As we move away from the tropics towards the poles, we lose habitat after habitat until we are left with only the icy desolation of the polar icecaps. At the same time, we lose opportunities for ecological speciation. This 'cradle effect' can also be seen in the aquatic environment, as suggested by the fact that most marine animal lineages, at least over the last 250 million years, began in tropical waters.

On the other hand, the tropics are also a museum of sorts because they define the latitude of the planet with the most stable temperatures over geological time. On this vast timescale, oscillations of the Earth's axis, variations in the Earth's orbit and changes in ocean currents caused by continental drift bring colder or warmer times, which mainly affect the poles and surrounding areas, as they are more sensitive to small alterations in the climate system. For example, where Antarctic mountains are buried under hundreds of metres of ice, there were once forests with polar dinosaurs. In what is now Norway, palm trees grew within the Arctic Circle 50 million years ago. At the equator, however, warm jungles have existed for practically as long as there have been forests on Earth. The thermal stability of the tropics must have reduced the risk of extinction of their species, as they have been subject to fewer relatively abrupt changes in temperature and vegetation over millions of years. We know, for instance, that these changes wiped out many forest species in the Mediterranean region over the last 10 million years. The stability of living conditions in the tropics must also have allowed speciation to occur there without interruption, thus reinforcing the 'cradle effect'.

We could go on and on disentangling the causes of the latitudinal gradient, as has been done for decades, but perhaps there would be little point, because we actually have very good reasons to expect diversity to increase towards the tropics on any inhabited planet more or less like the Earth, for the arguments above and probably for many others that are often cited in reviews of the gradient: the mid-domain effect (see glossary), the larger area occupied by the tropical climate zone compared to the temperate and cold climate bands (implying higher diversity in the species–area curve), and so on.

In general, therefore, any region like the tropics, any region that is stable and has a wide variety of habitats, will easily have high biodiversity. And we already know that this regional diversity spreads to local communities (see the end of Chapter 4), populating them with more species than in a poorer region. However, within a given region there are habitats with a lot of diversity, such as jungles, and others with very little, such as salt deserts or rocky peaks. What is the reason for this difference?

First, the area occupied by each habitat. If a habitat is small in area, there will be little room for species to originate in it. And those that do arise will be adapted to the particular conditions of that habitat, so will be largely confined to a small area, making them easy targets for extinction. This combination of difficult speciation and easy extinction leads to low diversity in habitats of small regional extent.

Furthermore, almost all species-poor habitats have conditions that are in some way extreme compared to the rest of their region (in terms of temperature, or drought, or salinity, etc.). So the typical inhabitants of the region will find it difficult to adapt to living there... And this is the problem with these extreme habitats: in order to have species, the ancestors of those species must have adapted to extreme conditions, which is difficult in itself – and this creates a vicious circle, or what is called a 'trap 22', which further promotes the poverty of such environments.

The area and trap-22 effects, first described by the ecologist John Terborgh, are very helpful in understanding the differences in diversity between habitats within a region, regardless of latitude. In addition, one can consider the availability of niches in each habitat, its disturbance regime, or the effect of predators. But none of these factors can add species to a particular habitat if evolution has not produced species capable of surviving there.

Complexity and diversity

The other big question about biodiversity is why some groups of organisms have more species than others. If the difference in the number of species is small, it may simply be by chance, because the evolution of biodiversity does not follow rigid rules but trends, with room for accident and chance, as happens in human history.

Leaving aside mere chance, there are many, many confirmed reasons why one group might be more diverse than another, all other things being equal. These include, for example, the age of the lineage (it takes time to produce many species), the area it occupies (the larger the area, the more opportunities for new species to emerge), body size (the world is bigger for smaller animals than for larger ones, and therefore also the effective area in which to speciate), sexual selection (if intense, it can facilitate speciation), dispersal ability (in principle, it makes it harder for new species to arise because it makes genetic isolation difficult), the length of generations (if they are short, evolution will be faster than if they are very long), the type of diet (since on land there are more plants than prey, new herbivore species are more likely to survive than carnivores), and a long etcetera. Some of these elements are important for certain groups but not for others, depending on the vagaries of biology, for reasons that would have to be ascertained on a case-by-case basis.

Well, is that all? Is there not one big theme unifying this whole issue? Yes, there seems to be, because of all the myriad causes of group diversity, one emerges as the most basic in the game: complexity.

If we think of an animal as complex as an insect, evolution can modify many, many things to adapt it: legs, wings, mouthparts, sensory organs, diet, complicated behaviour... All these possibilities disappear in an animal that is simpler, such as a coral. So should we be surprised that insects, or vertebrates, have many more species than corals or foraminifera? Probably not very surprised. The key point is that a complex organism can be modified by evolution to produce many variations that can adapt to very different habitats and ways of living. In comparison, a simple organism offers less scope for evolution and therefore less potential for diversity.

The importance of complexity as a breeding ground for diversity in fact follows from the three rules of macroevolution. These are three trends discovered by the palaeontologist Steven Stanley in the late 1970s which, despite their importance, have gone largely unnoticed by biologists in general. Stanley deduced them from the fossil record of animals, but they also seem to apply to plants. The three rules certainly deserve to be better known among friends of the game of species; let's get to know them:

> **First rule of macroevolution.** The more easily a group produces species, the greater the risk of extinction of its species (Figure 7.2). This implies that the higher the rate of speciation of a group of organisms, the higher its rate of extinction. This is probably because speciation rate mostly depends on the same causes as extinction rate, which is curious – we will come back to this in the third rule.
>
> **Second rule of macroevolution.** The higher the rate of speciation and extinction of a group, the faster it will produce species overall. This is because the rate of net species production, or rate of diversification, is the rate of speciation minus the rate of extinction. So if there were, say, 0.7 new species and 0.2 extinctions per species per million years, the rate of diversification would be 0.5; but if they were twice as fast, 1.4 and 0.4 (as one goes up, the other goes up in the same proportion, because they are roughly correlated according to the first rule), then the difference would be 1.0 – and so the increase in the net number of species would be twice as fast.
>
> **Third rule of macroevolution.** The most complex organisms tend to produce species very quickly. They readily give rise to species which also easily go extinct, so their net production of species is usually high. Perhaps this is because their complexity makes them very dependent on details of the environment, and so they evolve rapidly, leading to specialised forms that are very vulnerable to extinction when the environment changes.

According to the three rules of macroevolution, we would expect organisms of greater complexity to have more species. This is clearly confirmed by the comparisons shown in Table 7.1. The more complex groups always show much more diversity than their simpler closest relatives. This supports the idea that the link between complexity and diversity is one of the keys in the game of species. Accordingly, on any inhabited planet, we would expect the greatest diversity to be found in the most complex beings, not the simplest.

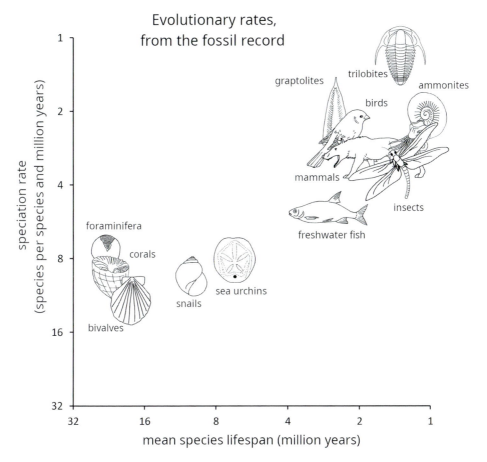

FIGURE 7.2 The relationship between speciation and extinction rates. The higher a group appears on the vertical axis, the easier it is for one of its species to give rise to a new species. The further to the right a group is on the horizontal axis, the more vulnerable its species are to extinction. As can be seen, the rate of speciation tends to increase with the rate of extinction. Interestingly, more complex animals usually have higher rates of evolution, if we take complexity in the broadest sense – trilobites and ammonites, with their complex sensory organs, ability to move quickly and active lifestyles, would be more complex than corals or clams. Based on a diagram by Steven Stanley (1990), with the drawings placed approximately at the midpoint of the rhythms of each group. Speciation rate refers to an evolutionary radiation.

Game manual

The rules of the game of the species are not inexorable laws like those of physics, but tendencies that always leave room for exceptions. Nothing more than trends, but also nothing less. In the following points I have tried to distil them, in the manner of a game manual:

TABLE 7.1 The greater the complexity, the greater the diversity. This table compares complexity and diversity in pairs of sister groups, that is, two lineages that are each other's closest evolutionary relatives, both descended from the same common ancestor and therefore separated from that ancestor by the same amount of time. I have included all the sister groups I have been able to find that clearly differ in complexity, measured as the number of cell types. The complexity and diversity shown here are reasonable estimates taken from the literature. As can be seen, in all cases the group of higher complexity has more species than the other. The difference is always more than one order of magnitude, so it does not depend on small details of the diversity estimates and is probably real. The probability of such a difference in these eight comparisons occurring by simple chance is almost negligible (<1%, assuming the probability of occurring and not occurring are the same). The conclusion seems to be that more complex organisms tend to be more diverse. For data sources, see the references for this chapter.

Sister groups	Complexity (number of types of cells)	Diversity (estimated number of species)	Is the most complex group the most diverse?
Phaeophyceae – brown algae	14	2,000	yes
Xanthophyceae – yellow-green algae	5	100	
Spermatophyta – seed plants	44	300,000	yes
Moniliformopses – horsetails and ferns	20	11,000	
Fungi – fungi	9	1,500,000	yes
Nucleariida – a group of protists	3	20	
Echinodermata – echinoderms	41	7,000	yes
Hemichordata – hemichordates	25	100	
Craniata – vertebrates and hagfishes	38–215	66,000	yes
Urochordata – ascidians (tunicates)	38	2,800	
Brachiopoda – brachiopods	34	335	yes
Phoronida – horseshoe worms	23	20	
Nematoda – roundworms	14	500,000	yes
Nematomorpha – horsehair worms	8	400	
Arthropoda – arthropods	42–90	>5,000,000	yes
Onychophora – velvet worms	30	180	

1. In nature the struggle for existence is the rule for living organisms, and this is why natural selection causes them to evolve and thus adapt.

2. Organisms that reproduce sexually form entities which are more or less genetically closed; we call them species and they are evolutionary units.

3. Most species arise through adaptation to a new environment or way of life, often involving geographical isolation.

4. Two species can compete with each other until one of them disappears from the ecosystem.

5. Natural selection tends to separate species into distinct ecological niches, thus preventing them from competing too much with each other.

6. In a given ecosystem, predators or environmental disturbances can maintain species that would otherwise be eliminated by competition.

7. The larger the area we explore, the more species there will be, as we will include more individuals, more habitats, and even more evolutionary regions.

8. The more species there are in a region, the more diverse their local communities tend to be.

9. In island-sized areas, the number of species tends to depend strongly on the balance between arrivals and extinctions.

10. Species expand their ecological niches in the absence of competitors, but contract them as competitors accumulate.

11. The more species there are, the harder it is to add another similar species, either by immigration or by speciation, because competition is stronger.

12. In continent-sized areas, there is a trend towards equilibrium between species that arise by speciation and species that go extinct.

13. Each species needs to evolve in order to survive, as its environment changes over time and evolution perfects its enemies.

14. Throughout the history of life, diversity has increased mainly because evolution has added species with different lifestyles to communities.

15. The history of life shows that biodiversity has gone through ups and downs linked to changes in the functioning of the Earth (climate, sea level, several catastrophes, etc.).

16. Tropical latitudes have the greatest biodiversity, because they have factors that favour speciation and make extinction less likely.

17. In a given region, most species are found in the extensive habitats with moderate conditions, not in the smaller and more extreme ones.

18. The number of species in a group can vary for many reasons, but in general more complex organisms tend to be more diverse.

After this condensed summary, now that we know the basics of how biodiversity works, the next chapter will look at what we humans are doing to this living heritage. We seem to have forgotten that we belong to it, and yet we depend on it to an extent that we need to recognise. If we continue to act as we have in recent centuries, what will be left of the game of species for future generations?

Chapter 8
The future of life

We share planet Earth with approximately five million species, perhaps three to eight million. But these numbers will not hold up for much longer if we continue to extinguish species at the current rate. It is a sobering fact that the rate of extinction today is at least hundreds of times higher, and probably thousands, than its average in the history of life, according to very conservative estimates by the International Union for Conservation of Nature (IUCN). To give a more accurate picture of the situation, the following data come from the summary statistics of the IUCN Red List.

A total of 926 extinctions have been recorded, and 81 species survive only in captivity but are extinct in the wild. These 1,007 extinctions are certainly an underestimate, as the vast majority of species are not monitored and we have a profound lack of knowledge about the biodiversity of many regions, especially those suffering the greatest losses, in the tropics. Apart from these confirmed extinctions, there are 1,346 species that may have disappeared, but where more information is needed to confirm their extinction. It is estimated that many species must have gone extinct without us even knowing they existed, as we still have a lot of biodiversity to discover.

Extinction has historically been most intense on islands, but in the last decades many extinctions have been recorded on the continents. We do not know the situation in freshwaters, but a large proportion of Europe's endemic freshwater fish are threatened. And the oceans, once considered more resilient to species loss, have proved to be less so.

The data paint a picture of a planet where probably one in three species, or more, is heading towards extinction. This proportion is derived from the IUCN's analysis of around 8% of all described species. That is approximately 166,000 species, of which the IUCN Red List records 46,337 as threatened with extinction to varying degrees, or about 28%. The situation for the few groups whose biodiversity we know reasonably well is alarming: the percentage of threatened species is 12% for birds, 23% for mammals and 36% for amphibians. Among plants, only gymnosperms are fully assessed, with 42% of species threatened. Over recent decades, trends in birds and amphibians indicate that threatened

THE FUTURE OF LIFE   67

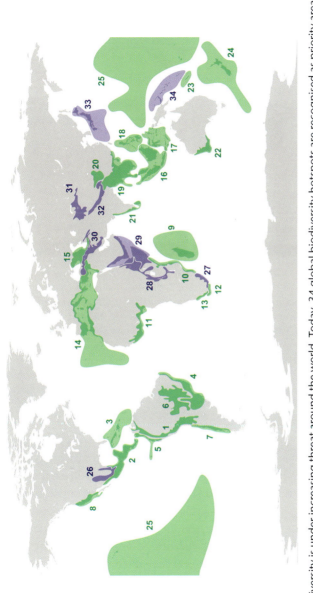

FIGURE 8.1 Biodiversity is under increasing threat around the world. Today, 34 global biodiversity hotspots are recognised as priority areas for species conservation. They stand out for their biodiversity but also for the degree to which they are threatened. These hotspots are home to half of all known plant species and more than three-quarters of all vertebrate species. (1) The Tropical Andes; (2) Mesoamerica; (3) The Caribbean Islands; (4) The Atlantic Forest; (5) Tumbes-Chocó-Magdalena; (6) The Cerrado; (7) Chilean Winter Rainfall–Valdivian Forests; (8) The California Floristic Province; (9) Madagascar and the Indian Ocean Islands; (10) The Coastal Forests of Eastern Africa; (11) The Guinean Forests of West Africa; (12) The Cape Floristic Region; (13) The Succulent Karoo; (14) The Mediterranean Basin; (15) The Caucasus; (16) Sundaland; (17) Wallacea; (18) The Philippines; (19) Indo-Burma; (20) The Mountains of Southwest China; (21) Western Ghats and Sri Lanka; (22) Southwest Australia; (23) New Caledonia; (24) New Zealand; (25) Polynesia and Micronesia; (26) The Madrean Pine–Oak Woodlands; (27) Maputaland–Pondoland–Albany; (28) The Eastern Afromontane; (29) The Horn of Africa; (30) The Irano-Anatolian; (31) The Mountains of Central Asia; (32) Eastern Himalaya; (33) Japan; (34) East Melanesian Islands. In purple, recent additions. Map attributed to Crates.

FIGURE 8.2 Some of the species that have become extinct due to human activity in recent centuries. From left to right, and from top to bottom: Thylacine *Thylacinus cynocephalus*, the largest carnivorous marsupial in historic times, extinct by the 1930s; Passenger Pigeon *Ectopistes migratorius*, once the most abundant bird on the planet, hunted until it had disappeared from the wild by the end of the nineteenth century; Great Auk *Pinguinus impennis*, a large North Atlantic seabird that became extinct in the mid-nineteenth century due to uncontrolled hunting coupled with egg collection and other factors; Dodo *Raphus cucullatus*, a huge member of the pigeon family, endemic to Mauritius and exterminated by humans in the late seventeenth century. Images by Baker & E.J. Keller, Hayashi & Toda, John James Audubon and Roelant Savery, respectively.

species are generally experiencing a worsening of their survival prospects, and the same is suggested by the limited information available for other groups. Disappearing populations and declining abundance are the order of the day for many species. But what are we doing? How did we get here?

There is no doubt that humans have caused the vast majority of extinctions recorded in history. Indeed, for tens of thousands of years, the advance of humans has been closely linked to the disappearance of the largest animals in each region, the so-called Pleistocene megafauna. Only in Africa has much of the large animal fauna survived, perhaps because it was there that humans evolved, giving African megafauna time to adapt to the progressive development of our hunting skills. For the last few centuries, the IUCN report identifies three main causes of extinction related to human impact: habitat loss, invasive species, and direct persecution, the three horsemen of the species apocalypse. In addition, diseases are becoming increasingly important as agents of species extermination, and the climate change we are causing is emerging as a major threat to biodiversity.

Habitat destruction

It is not difficult to understand that habitat destruction and degradation can easily wipe out many species in a world where we have converted more than 35–40% of the area once occupied by forests and other natural vegetation to cropland and pasture. This is a general figure, but in the case of the Mediterranean forest the percentage is around 70%. Logically, organisms that specialise in a single habitat type are worse off in the face of this loss, while generalists have other cards to play. Similarly, widely distributed species face lower risks than very restricted endemics, because, for many of the latter, the destruction of a few square kilometres of habitat can mean extinction. In this way, we are eliminating much of what makes nature unique in each region. We are leaving behind a planet of generalists with large geographical ranges, the same situation that tends to follow the mass extinctions that have plagued the history of life (see *The balance of evolution* in Chapter 6). It will be a good world for those who love rats, foxes and cats, but not for those who appreciate the value of rare and unique species, such as many large animals, freshwater fish and tropical or mountain flowers.

Habitat destruction tends to leave isolated patches of nature within a matrix of grassland and crops. This process, called fragmentation, creates 'islands' of habitat in an 'ocean' of adverse conditions. And we already know from Chapter 5 what happens to biodiversity in these fragments: as if they were real islands, they lose species until they reach a lower diversity balance than before fragmentation. Thus the impoverishment typical of islands leads to the disappearance of many species in fragmented landscapes. In these landscapes we try to stop the devastation by creating protected areas, larger 'islands' that would gradually lose their biodiversity, especially as we continue to destroy it around them instead of promoting its recovery.

Many people think that conserving biodiversity means creating protected areas, but species must also be actively protected outside these nature reserves. This is an obligation under Article 8 (d and e) of the most important international agreement for nature conservation, the Convention on Biological Diversity, a United Nations treaty to which we will return at the end of this chapter. On the contrary, too often the creation of reserves seems to be used by governments as an excuse to allow actions that accelerate the degradation of nature in the unprotected matrix surrounding these areas. In doing so, we increasingly isolate their inhabitants, thus facilitating their loss through the inexorable mechanism of island impoverishment.

Invasive species

Another must contained in Article 8 of the Convention (point h) is the control of alien species, better known as invasive species, which can cause serious losses both to the living heritage and to the economy. A species introduced by humans, either accidentally or deliberately, into an area outside its natural range may harm other species as a competitor, parasite or predator. For instance, competition with the Dingo, the subspecies of wolf introduced to Australia by the Aboriginal inhabitants, probably drove the Thylacine (Figure 8.2) to extinction there before the first Western settlers arrived.

Perhaps the Dingo also spread disease to the Thylacine, but to consider this risk in more detail, let us visit Hawaii one more time. In 1826, the Southern House Mosquito *Culex quinquefasciatus* was unintentionally introduced to the island of Maui in barrels of water brought by the sailing ship *Wellington*. Until then, there were no blood-sucking mosquitoes in the archipelago, so migratory birds arriving from the mainland with certain diseases could not infect native birds, because there was no vector to carry the microbes from the blood of one bird to another. But since its arrival, *Culex* spread rapidly, and as a result bird flu, malaria and other diseases decimated Hawaiian birdlife, causing die-offs that contributed to the extinction of at least 10 species.

The danger of invaders is heightened when they are predators introduced into communities that lack this type of natural enemy. The newly arrived predators will find naive prey who do not know they are dangerous, or how to avoid or defend themselves well. This can lead to real species massacres, such as that caused by 'the snake that ate Guam'. This is the title of an article by ecologist Stuart Pimm on the disastrous consequences of the arrival of the Brown Tree Snake *Boiga irregularis* on the Pacific island of Guam. The reptile, which is native to Australia, the Solomon Islands, Indonesia and Papua New Guinea, probably arrived as a stowaway aboard a military aircraft in the late 1940s or early 1950s. In the absence of natural enemies, it spread across the island, and wherever it went, bird numbers plummeted within a few years. Experiments on Guam showed the snake's enormous capacity to catch island birds. It is credited with the extinction of 12 of Guam's 22 bird species, and 3–5 of its 10–12 reptile species.

Guam is just one of the many examples of how vulnerable island species are, as they are not adapted to living with many competitors, predators or parasites, due to the low diversity typical of islands. This is why common animals such as cats and rats can devastate the fauna of entire islands. In fact, both the domestic cat and the Black Rat *Rattus rattus* are included in the IUCN's list of the 100 most damaging invasive alien species in the world. This document reveals nonsenses such as the introduction of the Rosy Wolfsnail *Euglandina rosea*, a snail predator, since the 1950s to control agricultural plagues of the Giant African Land Snail *Achatina fulica* on many Pacific and Indian Ocean islands, where the latter had

FIGURE 8.3 Some of the 100 most damaging invasive alien species in the world, according to the IUCN. Above left, Brown Tree Snake *Boiga irregularis*, which devastated birdlife on the island of Guam. Above right, Common Water Hyacinth *Eichhornia crassipes*, an ornamental plant that has become naturalised on five continents and is capable of doubling its population in little more than 12 days, soon filling rivers and damaging native species. Below left, European Rabbits *Oryctolagus cuniculus* in a myxomatosis experiment on Wardang Island, near Australia, in 1938. In an attempt to control plagues of European Rabbit in Australia, the rabbits were inoculated with myxomatosis from an American rabbit species, but through a fatal chain of blunders the virus eventually decimated European Rabbit populations in Europe. Below right, a Northern Pacific Sea Star *Asterias amurensis*, which has invaded the coasts of several countries through its larvae infiltrating the waste tanks of ships. Photographs by Pavel Kirillov, H. Zell, National Archives of Australia and perhaps Lycoo, respectively.

been introduced for human consumption. But the Rosy Wolfsnail itself became another problem, rapidly wiping out endemic snail species on these islands; in French Polynesia, for example, it is threatening many species of tree snails of the genus *Partula*. So it does not seem a good idea to try to eliminate an invasive species by adding another species that may cause further damage.

Ever since the ecologist Charles Elton first brought the global problem of biological invasions to light in his book *The Ecology of Invasions by Animals and Plants* (1958), huge sums of money have been spent trying to control the economic and environmental damage caused by the hundreds of species that have been the protagonists of these 'biological explosions', as he metaphorically called them.

In 1996, the ecologists Mark Williamson and Alastair Fitter proposed three rules of thumb for assessing the risk of these impacts: of all the species that humans bring into a region, about 10% will escape human control and go wild; of these, about 10% will establish populations and become introduced species; and of these, about 10% will cause problems and be called pests. The data they reviewed to evaluate these '10% rules' only support them in about half the cases, but when they fail it is almost always because they underestimate the true percentage. Even so, these rules can provide useful insights. For example, invasive birds in Hawaii seemed to be an exception, because of their enormous success as invaders, but they do meet the 10% rule if we consider only those that have invaded native island habitats, not lowland 'invasive' habitats artificially expanded by humans. This reveals one of the reasons for the success of these birds, namely that many of them occupy human-modified habitat types that are currently expanding.

Direct persecution

The third horseman of the current species apocalypse, direct persecution, can wipe out even extraordinarily abundant animals. This is demonstrated by the decline of the Passenger Pigeon (Figure 8.2), a North American species that was probably the most abundant bird in the world in recent centuries. In the early nineteenth century, the ornithologist John James Audubon witnessed how migrating flocks of this species darkened the sky as in an eclipse of the sun. Based on these flocks, several authors estimated the number of birds to be between one and three billion. If these figures are not correct, it is probably because they are too low. But the arrival of firearms in the United States, together with the great advance of deforestation there, sealed the fate of *Ectopistes migratorius*. When a flock flew overhead, someone armed with a shotgun could easily kill six pigeons with a single shot, simply by aiming at the sky. These pigeons were killed in their millions, without a care in the world, for how could an animal capable of such astronomical numbers become extinct? And yet the last Passenger Pigeon, named Martha, died in Cincinnati Zoo on 1 September 1914. There are many more examples of species hunted to extinction, but I think none is more starkly depressing than the case of this pigeon.

Other causes of extinction

On top of these threats we have added climate change, which increased global surface temperature in 2011–2020 to 1.1 °C above 1850–1900 values, according

THE FUTURE OF LIFE   73

to the 2023 report of the Intergovernmental Panel on Climate Change (IPCC). This rise may seem small, but it is increasing and already means significant temperature differences for several months of the year in many ecosystems.

In the IUCN data, there is growing evidence that climate change will become one of the main drivers of extinctions during the twenty-first century. It works by altering or eliminating habitats, transforming them into environments dominated by species better adapted to the new higher temperatures. Global warming is now seriously affecting not only the Arctic, where melting ice is leaving the Polar Bear, the flagship of this problem, with nowhere to hunt, but also and especially the tropics, where many forms of life are highly sensitive to changes in temperature. The IUCN estimates that 35% of birds, 52% of amphibians and 71% of corals have characteristics that make them vulnerable to climate change, with 70–80% of species in these groups classified as threatened. The situation is most worrying for corals, which have declined dramatically in recent decades. The loss of corals is doubly serious because the reefs they build support the greatest biodiversity in the marine environment. Amphibians are

FIGURE 8.4 From left to right, and from top to bottom: Polar Bear *Ursus maritimus*, one of the species most threatened by global warming; corals of the Great Barrier Reef of Australia, an ecosystem highly sensitive to climate change; Golden Toad *Incilius periglenes*, endemic to Costa Rica and emblematic of the mass extinction of amphibians, as it has not been seen since 1989 and is now classified as extinct by IUCN; Iberian Lynx *Lynx pardinus*, endangered by illegal hunting, trampling and other factors, but now with better prospects thanks to captive breeding programmes. Photographs by Alan D. Wilson, Toby Hudson, Charles H. Smith and the Iberian Lynx Ex-Situ Conservation Programme, respectively.

also disappearing rapidly, owing to a number of factors including not only their sensitivity to climate change, but also pollution and the spread of the lethal disease chytridiomycosis.

Like a chain of dominoes, each species that goes extinct usually brings down several more. For example, when the plant Eelgrass *Zostera marina* disappeared from large parts of the North Atlantic, possibly because of a disease that decimated it between 1930 and 1933, its decline led to the extinction of the small snail *Lottia alveus*, which was once abundant along the east coast of North America, but which had the misfortune to feed only on this plant and to live only in that region. With each species that disappears we may also lose others that depend on it: its exclusive herbivores, its specialised parasites, the predators whose diets are eventually based on it, or the symbionts that need it to survive.

This ominous domino effect, known as 'extinction cascade', means that the current level of threat to biodiversity is underestimated. In 2004, Koh Lian Pin and his colleagues calculated the average number of species which are closely dependent on each individual species by analysing various parasite–host, pollinator–plant and commensal–host systems. They calculated that around 6,300 species must be threatened by the decline of other species already listed by the IUCN. This would increase the number of endangered species by about a third of the IUCN estimate.

Why does it matter?

But why should we care about species loss? Nature lovers need not even ask this question, but to be effective in defending biodiversity we need to arm ourselves with reasons. We can find a good list of them in the major international treaty that legally underpins its protection, the Convention on Biological Diversity. Created by the United Nations and signed by most of the world's countries in Río de Janeiro on 5 June 1992, it has been in force since 29 December 1993 and continues to be applied today. The preamble of the Convention begins by making clear why we should care about biodiversity, quickly listing its many values:

> Conscious of the intrinsic value of biological diversity and of the ecological, genetic, social, economic, scientific, educational, cultural, recreational and aesthetic values of biological diversity and its components…

Let's pause to unpack some of these values.

The intrinsic value would be the deepest of all, for it means that every species is valuable simply because it exists. We value the works of our own species, our monuments, our art; so how much should we value every other species, which we did not make, which have been given to us as the fruit of billions of years of evolution? On the other hand, just as the value of a painting is not in the amount of wall space it covers, so the value of a species does not lie in the role it plays in nature. And just as monuments make a country unique, so its endemic species make it unrepeatable. All these considerations justify that biodiversity has a value in itself.

The role of each species in nature must also be valued, and this brings ecological values to the fore. Species in nature can produce food, recycle nutrients, control potential pests – activities that are essential to maintaining biodiversity. These functional values are often mixed with economic benefits, as species provide us with many useful ecological services that are very difficult to include in standard macroeconomic indicators. How much should we pay for the oxygenation of the air, the pollination of crops, the regulation of potential pests by their natural enemies, the recycling of dead matter by decomposers, the maintenance of the soil by roots to prevent erosion? How can all these ecological services, which are essential to humanity, be included in indicators such as gross domestic product (GDP)?

In considering the economic value of diversity, we sometimes lose sight of the fact that we feed on it, for the variety of animal and plant breeds that sustain us is also biodiversity. However, especially in recent centuries, we have selected only a handful of livestock breeds, vegetable varieties and fish species to feed us, ignoring many others that we should be exploring because they could provide us with food that is perhaps of better quality or easier to obtain. Moreover, traditional livestock and crop varieties, adapted to the conditions of each region, are a heritage that should not be lost, but rather enhanced.

In addition to food, biodiversity also provides us with raw materials (fibres for basketry, cork, wood, etc.) and many medicines. Vinblastine and vincristine, two substances used to treat some cancers, are extracted from the Madagascar Periwinkle *Catharanthus roseus*, a beautiful plant that, like so many others, has developed chemical defences against herbivores over millions of years, which now happen to be useful in medicine. Many of the medicines we use are derived, directly or indirectly, from plants and other organisms, and the presence of possible pharmacological principles remains to be investigated in most species.

Given all this wealth of values, we can only conclude that biodiversity is a heritage that must be protected so that humanity can continue to enjoy and benefit from it. In this sense, the purpose of the Convention on Biological Diversity is succinctly stated in its first article:

> The objectives of this Convention … are the conservation of biological diversity, the sustainable use of its components and the fair and equitable sharing of the benefits arising out of the utilisation of genetic resources.

In expressing it in this way, the UN and the signatory countries made an explicit link between nature conservation and sustainable development in the pursuit of a more equitable social order on an international scale. This was the right approach, because protecting the game of species cannot be separated from human development and the economy. Nature is so intertwined with the human enterprise that if we are to succeed in conserving wildlife, we must change many aspects of our current global society. We can't just sit back and wait for better times, because the rate of biodiversity loss is not only rapid, it's accelerating.

These goals of the Convention were crystallised in the UN's Project XXI to promote sustainable development, approved at the same Rio Summit in 1992 and implemented through local Agenda 21 programmes. However, decades

FIGURE 8.5 Species whose chances of survival have been greatly improved by conservation efforts. Above, the Grey Whale *Eschrichtius robustus*. Below, two endangered species that disappeared from the wild, with a few individuals remaining in captivity, and are now being reintroduced through breeding programmes: left, Black-footed Ferret *Mustela nigripes* (North America);right, Przewalski's Horse *Equus ferus przewalskii* (Mongolia, the only remaining subspecies of the wild horse). Photos by Merrill Gosho, J. Michael Lockhart and Bouette, respectively.

after its inception, this project has not succeeded in transforming society by establishing a truly sustainable culture, because this task requires a willingness to change that does not yet exist in many sectors. It is not a question of choosing one political idea or another, but rather that if we continue to do the same things to the planet, then the same things will surely continue to happen, so if we want to achieve something different, we need to change our actions.

Similarly, most of the countries that signed the Convention are at best only superficially complying with it, and while there have been some conservation successes, overall the problems for species have been exacerbated rather than solved, as the IUCN reports clearly show. It is fashionable in many governments to endorse sustainability and conservation, and we appease our consciences by celebrating Earth Days, Biodiversity Days, and various other special days, but meanwhile we as a society continue to destroy nature and in practice do nothing to remedy this except in very specific cases. There is hardly any investment in actively protecting biodiversity inside and outside protected areas, while there is plenty of money for other, less productive purposes. Instead of the principle of prudence that the Convention recommends should guide our actions in ecosystems when we are in danger of damaging them, the short-term maximum benefit is often sought, at any cost. Paradoxically, contradictions are everywhere, but so is the will to resolve them. So what is wrong?

What kind of future can we expect for life? The question is not whether we will become extinct if we carry on as we are, but what kind of world we want for ourselves and for future generations: one full of living heritage and resources, or one so impoverished that every walk in nature reminds us of how incompetent we humans have been in caring for our own home. Nor is it a choice between welfare and conservation, because we know enough about technology and ecology to reconcile the two. For instance, solar photovoltaic panels on roofs could provide much of the energy we need, freeing us from fossil fuels and thus reducing global warming, if electricity companies and governments would facilitate their installation and development. This is just one of the many contradictions that could be resolved if we prioritised what can truly improve the not-so-distant future of humanity.

We are already in the midst of a mass extinction, and the consequences are here, but we can still avoid even greater losses. If they do occur, life will need at best several million years to recover, assuming that species production can be reactivated normally in the future world we humans are creating, which is doubtful. In any case, the examples of species and ecosystems that have been saved from extinction give us a glimmer of hope. Something can be done, even against the tide.

***

So I close this book, dear reader, by thanking you for having accompanied me on this adventure of exploring the intricacies of the game of species, for having travelled the world with me, with the freedom of imagination to discover so many wonderful natural stories. I hope these pages have been of help in appreciating the vast game that life plays around us, that unifies oceans and mountains across millions of years, always following the same rules.

Ideas, like the species on the Red Queen's board, must always be on the move to avoid extinction or being locked away in an ivory tower, far from society, as has happened to many of those discussed in this book. So please do not hesitate to share the ideas, especially from this last chapter: that safeguarding our living heritage is not just a whim, but rather a crucial responsibility of every generation; that preserving all the life that surrounds us is not only our ethical duty, but also a way of working for a fairer and more sustainable world. And since ideas are empty unless they are put into practice, I end by inviting you to do something about the mass extinction we are witnessing right now, so that the game of species will continue to be played, and the richness of the spectacle of life on Earth will continue to fascinate humanity for many, many generations to come.

Glossary

Including some terms which, for one reason or another, have been omitted from the main text but deserve to be included here.

Adaptation: in an organism, an inherited characteristic that improves the chances of survival, successful reproduction, or both. It is the result of natural selection.
Allopatric speciation: speciation associated with the geographical isolation of a population from other members of the species from which it evolved.
Alpha diversity: local diversity, that is, the biodiversity of a community, on a small scale. For simplicity, the terms alpha, beta and gamma diversity are not used in this book.
Beta diversity: diversity between communities, that is, differences in the species composition of different communities.
Biodiversity: the variety of life at all levels, from genes to ecosystems.
Biogeography: the study of the geographical distribution of living things.
Coevolution: evolution due to the interaction of two types of organisms. For example, much of the diversity of insects is due to insect–plant coevolution: there are many species of beetles and butterflies that have specialised in eating the leaves of a single type of plant, and so as new plant species appear, new insect species evolve by adapting to feed on those plants. Thus the evolution of insects and plants is intertwined.
Community: a group of species that coexist in an ecosystem. In this book the term 'community' refers only to biodiversity on a small scale, say from less than one hectare to a few hundred square kilometres (this is local diversity, or biodiversity within a particular habitat).
Competitive exclusion: extinction of one species due to competition from another.
Court Jester: the hypothesis that biodiversity throughout the history of life has been highly dependent on physical variations in the Earth system, such as sea-level rise or fall and climate change.
Diversity: in general, biodiversity; in this book, the term often refers to the number of species (richness).
Ecological release: expansion of a species' niche in the absence of competitors. It usually occurs on islands (because they are species-poor compared to areas of

the same size on the nearest continent) and also after mass extinctions (because few species remain).

Ecological speciation: speciation by adaptation to ecological conditions different from those preferred by the ancestor.

Ecosystem: a set of organisms and their inert environment, in which living things interact with each other and with the non-living components.

Endemic: a type of organism that exists only in a particular region. A species may be endemic to a valley, or to a country or to a continent.

Eukaryote: an organism whose cells have a nucleus (or nuclei) containing DNA. A eukaryotic cell is usually larger and internally much more complex than a prokaryotic cell. Animals, fungi, plants, algae and protists are eukaryotes.

Evolution: change in the hereditary characteristics of populations of living organisms over generations.

Evolutionary radiation: rapid production of species over a short time span in geological terms, typically a few million years. It usually occurs in situations of low diversity: when a desert island is colonised, after a mass extinction, or when an organism emerges with such a novel way of life that it has little competition.

Expansive equilibrium: slowed growth of biodiversity on a regional scale and over geological time. The number of species increases as evolution opens up new ways of life, but in these new niches the rise of diversity is slowed by the negative interactions between species.

Gamma diversity: regional diversity.

Generalist: an organism whose ecological niche is broad compared to that of another, which would be a specialist.

Geological time: time on the scale of Earth history, measured in millions of years.

Gradualism: classical view of evolution in which the origin of species occurs gradually, without acceleration as in punctuated equilibrium. Darwin adopted gradualism because it seemed more reasonable as a premise for a scientific theory, without clear evidence that it was actually true. Punctuated equilibrium seems to fit best with most of the evolutionary histories we now know.

Habitat: the place where an organism lives; more broadly, the type of environment in which certain species tend to live.

Hybridogenesis: formation of a new species by hybridisation between two closely related species, resulting in a stable, reproductive hybrid. This sometimes happens, but because the two parent species can hybridise it raises concerns about their status as valid species.

Keystone predation: phenomenon whereby predators increase the diversity of their prey by consuming species that compete with them, thus allowing less competitive species that would otherwise be exterminated by the former to remain in the community.

Instantaneous speciation: speciation by a sudden genetic change, usually by multi-plication of chromosome number, occurring in a short interval of time, on the order of the lifespan of an individual or less.

Intermediate disturbance: hypothesis that diversity peaks when small catastrophes, called disturbances, occur in the ecosystem at an intermediate frequency, preventing the more competitive species from exterminating the weaker ones. The intermediate disturbance hypothesis provides a good explanation for the biodiversity of certain small-scale ecosystems.

Island biogeography: the idea that the balance between immigration and extinction of species determines the diversity of an area the size of a landscape, a province, an island or a small country, with the result that a stable equilibrium in the number of species is usually reached.

Latitudinal gradient of biodiversity: the tendency for diversity to increase with decreasing latitude. It is one of the two most fundamental patterns of biodiversity, along with the species–area curve.

Limiting similarity: the maximum similarity that two species can have without one eliminating the other through competition. It is doubtful that such a threshold really exists.

Macroecology: statistical study of the ecological characteristics of a large number of species, usually on a large scale (continents, oceans). The macroecological focus is on traits such as body size, abundance, diversity, geographic range size, etc.

Macroevolution: evolution viewed on a timescale where it is mainly manifested by the birth and extinction of species.

Mass extinction: an event in the history of life in which approximately 50%, and often more, of the species on the planet become extinct.

Metabolic theory of ecology: theory linking different ecological and evolutionary patterns (body size, relative abundance of species, birth and death rates, diversity, etc.) to the physical equations that describe metabolism in physical terms.

Mid-domain effect: the accumulation of species in the middle part of a geographical area because this is where the geographical ranges of species tend to overlap. To understand this, imagine a metal case (continent) with many pencils of different lengths (geographic ranges of species) inside. If we shake it randomly, when we open it we will almost always find more pencils overlapping in the middle of the case (lengthwise) than at the ends. This effect may help explain why the tropics are so diverse.

Natural selection: the process of differential reproduction that gives rise to adaptations. Natural selection occurs when a heritable trait increases or decreases the probability that the individuals possessing it will survive and/or reproduce successfully.

Niche: the position occupied by a species in nature, defined by its relationships with other species and with the inert environment.

Niche differentiation: the process by which competition between two species triggers natural selection to reduce the overlap of their niches, thereby reducing the disadvantage that the species suffered from competing against each other. Also known as niche segregation. It causes partitioning of resources between species in a community.

Paradox of the plankton: the apparent contradiction between the principle that each species occupies a single niche and the large number of species that tend to occur in the plankton, the community of tiny organisms floating in water. In 1961, Hutchinson noted that waters seem to offer very few niches to occupy, because the physical and chemical conditions are very uniform due to the mixing of water masses by currents. This scarcity of niches should mean few species, but in fact the opposite is true, as plankton diversity is often enormous. Possible solutions to this paradox have been proposed, mechanisms that reduce competition between species in the plankton and allow for high diversity despite the limited niches available. Such solutions include keystone predation by the many plankton-eating creatures, intermediate disturbance caused by frequent changes in temperature, chemical composition, etc. in surface waters, and the confirmed existence of aquatic microenvironments even in apparently well-mixed waters, with different physicochemical conditions and a duration long enough to favour one species or another.

Parallel speciation: the independent origin of new species from the same ancestor in independent cases, under similar ecological conditions, resulting in the evolution of the same traits.

Peninsula effect: species impoverishment of a peninsula compared with an area of the same size on the landmass to which it belongs. This effect occurs because the peninsula only receives immigrant species from its connection to the landmass, resulting in low diversities in the immigration–extinction balance of island biogeography theory.

Peripatric speciation: allopatric speciation occurring at the periphery of the geographical distribution of the ancestral species.

Polyploidisation: multiplication of the number of chromosomes, a genetic accident that can give rise to new species, especially in plants.

Prokaryote: an organism whose cells do not have a nucleus containing its DNA. The typical prokaryotic cell is smaller and much simpler than a eukaryotic cell. Bacteria and archaea are the only known prokaryotes.

Punctuated equilibrium: hypothesis proposed by palaeontologists Stephen Jay Gould and Niles Eldredge in the 1970s, stating that most evolutionary change occurs during speciation and that speciation takes little time compared to the typical lifespan of species. There is much evidence in favour of punctuated equilibrium, although this does not mean that its opposite, gradualism, is not valid in some cases.

Red Queen: the idea that species are forced to evolve constantly because their environment deteriorates over time, because the abiotic conditions to which they are adapted change, or because evolution improves their competitors, predators and parasites.

Richness: number of species, a useful measure of biodiversity in general, since it tends to be a valid proxy for biodiversity at other levels, such as genetic and ecological.

Sexual selection: a type of natural selection in which the selected trait increases or decreases the attractiveness of the organism to the opposite sex, thus affecting its chances of reproduction.

Specialist: organism whose ecological niche is narrow compared to that of another, which would be a generalist.

Speciation: the process that gives rise to a new species.

Species: fundamental evolutionary unit. According to the biological concept of species, this would be a set of populations of the same life form that can reproduce with each other giving rise to fertile offspring, but not with other similar entities. However, it is usually observed that closely related species can interbreed in this way, and this hybridisation occurs more often than previously thought, sometimes resulting in new species. The biological species concept applies only to sexually reproducing organisms; asexual beings do not have comparable species, but are also classified into 'species' according to their morphological or genetic characteristics.

Species–area curve: graphical representation of how biodiversity increases with area. Its three sections, with different slopes, form one of the two most fundamental patterns of biodiversity, along with the latitudinal gradient. The species–area curve is often confused with its middle section, whose equation is usually represented as $S = cA^z$, where S is the number of species, c is a constant, A is the area, and z is a value ranging from 0.12 to 0.18 for continents and from 0.25 to 0.35 for archipelagos (if the points of the curve represent islands). In the initial and final sections, S would be approximately proportional to the area.

Species selection: a phenomenon of macroevolution whereby lineages with high speciation rates, or low extinction rates, or both, tend to survive over millions of years. It is analogous to natural selection, but with species rather than individuals.

Sympatric speciation: origin of a species without geographical isolation.

Unified neutral theory of biodiversity and biogeography: theory developed by the ecologist Stephen P. Hubbell in the early twenty-first century based on the assumption that all species in a community have the same birth and death rates. The neutral theory shocked ecology for a few years because it ignores any ecological differences between species, but makes good predictions about the species–area curve (its middle section) and patterns of relative abundance of species, among other things. However, subsequent studies have shown that the predictions of this theory are not as accurate as those of earlier and more realistic ecology-based models.

Vicariance: speciation that occurs when the range of an ancestral species is divided by the emergence of new geographical barriers (mountains, seas, etc.), resulting in distinct isolated populations result that eventually evolve into new species.

References and further reading

Chapter 1

A cornerstone in biodiversity conservation: Secretariat of the Convention on Biological Diversity (2011) *Convention on Biological Diversity: text and annexes*. United Nations Environment Programme. www.cbd.int

Number of species tends to be highly correlated to other measurements of biodiversity: Gaston, K.J. and Spicer, J.I. (2013) *Biodiversity: an Introduction*, 2nd edition. Chichester: Wiley-Blackwell.

Use of the term 'diversity' rather than 'richness' to refer to number of species: Rosenzweig, M.L. (1995) *Species Diversity in Space and Time*. Cambridge: Cambridge University Press. https://doi.org/10.1017/CBO9780511623387

Cactus finches on Genovesa Island: Grant, B.R. and Grant, P.R. (1989) *Evolutionary Dynamics of a Natural Population: the Large Cactus Finch of the Galapagos*. Chicago, IL: University of Chicago Press.

Quotations from Darwin: Darwin, C. (1859) *On the Origin of Species by Means of Natural Selection, or the Preservation of Favoured Races in the Struggle for Life*. London: John Murray. Most of the citations are from Chapter III, but the estimation of the spread of an elephant pair appears in Chapter II.

House Sparrows in New England: Bumpus, H.C. (1898) Eleventh lecture. The elimination of the unfit as illustrated by the introduced sparrow, *Passer domesticus* (A fourth contribution to the study of variation). *Biological Lectures from the Marine Biological Laboratory of Woods Hole*: 209–228.

Achromatopsia and the Pingelap atoll: Hussels, I.E. and Mortons, N.E. (1972) Pingelap and Mokil atolls: achromatopsia. *American Journal of Human Genetics* 24 (3): 304–409.

The case of the Peppered Moth:

– Majerus, M.E.N. (2009) Industrial melanism in the Peppered Moth, *Biston betularia*: an excellent teaching example of Darwinian evolution in action. *Evolution: Education and Outreach* 2: 63–74. https://doi.org/10.1007/s12052-008-0107-y

– Cook, L.M., Grant, B.S., Saccheri, I.J. and Mallet, J. (2012) Selective bird predation on the peppered moth: the last experiment of Michael Majerus. *Biology Letters* 8: 609–612. https://doi.org/10.1098/rsbl.2011.1136

Size and reproduction in Coho Salmon: Van den Berghe, E.P. and Gross, M.R. (1989) Natural selection resulting from female breeding competition in a Pacific salmon (Coho: *Oncorhynchus kisutch*). *Evolution* 43 (1): 125–140. https://doi.org/10.1111/j.1558-5646.1989.tb04212.x

Sexual selection: Krebs, J.R., Davies, N.B. and West, S.A. (2012) *An Introduction to Behavioural Ecology*, 4th edition, Chapter 7. Chichester: Wiley-Blackwell.

The rapid evolution of Guppies: Reznick, D.N., Shaw, F.H., Rodd, F.H. and Shaw, R.G. (1997) Evaluation of the rate of evolution in natural populations of Guppies (*Poecilia reticulata*). *Science* 275 (5308): 1934–1937. https://doi.org/10.1126/science.275.5308.1934

Chapter 2

The diversity of groups of organisms: IUCN Red List version 2024-2: Table 1a. www.iucnredlist.org/resources/summary-statistics

Origin of species by natural selection:
– Darwin, C. (1859) *On the Origin of Species by Means of Natural Selection, or the Preservation of Favoured Races in the Struggle for Life.* London: John Murray.
– Schluter, D. (2009) Evidence for ecological speciation and its alternative. *Science* 323: 737–741. https://doi.org/10.1126/science.1160006

Speciation by natural selection in sticklebacks:
– Schluter, D. and McPhail, J.D. (1992) Ecological character displacement and speciation in sticklebacks. *The American Naturalist* 140: 85–108. https://doi.org/10.1086/285404
– Schluter, D. (1994) Experimental evidence that competition promotes divergence in adaptive radiation. *Science* 266: 798–801. https://doi.org/10.1126/science.266.5186.798
– Rundle, H.D., Nagel, L., Boughman, J.W. and Schluter, D. (2000) Natural selection and parallel speciation in sympatric sticklebacks. *Science* 287: 306–308. https://doi.org/10.1126/science.287.5451.306

Satellite species in lampreys:
– Hubbs, C.L. (1940) Speciation of fishes. *The American Naturalist* 74: 194–211. https://doi.org/10.1086/280888
– Vladykov, V.M. and Kott, E. (1979) Satellite species among the Holarctic lampreys (Petromyzonidae). *Canadian Journal of Zoology* 57: 860–867. https://doi.org/10.1139/z79-106
– Mateus, C.S., Almeida, P.R., Mesquita, N., Quintella, B.R. and Alves, M.J. (2016) European lampreys: new insights on postglacial colonization, gene flow and speciation. *PLOS ONE* 11: e0148107. https://doi.org/10.1371/journal.pone.0148107

The two forms of the periwinkle:
– Butlin, R.K., Galindo, J. and Grahame, J.W. (2008) Sympatric, parapatric or allopatric: the most important way to classify speciation? *Philosophical Transactions of the Royal Society B* 363: 2997–3007. https://doi.org/10.1098/rstb.2008.0076
– Johannesson, K. (2001) Parallel speciation: a key to sympatric divergence. *Trends in Ecology and Evolution* 16: 148–153. https://doi.org/10.1016/S0169-5347(00)02078-4

Divergence in *Timema* stick-insects: Nosil, P. and Sandoval, C.P. (2008) Ecological niche dimensionality and the evolutionary diversification of stick insects. *PLOS ONE* 3: e1907 https://doi.org/10.1371/journal.pone.0001907

Sexual selection matters in the origin of species but does not seem to be the main triggering mechanism:
- Ritchie, M.G. (2007) Sexual selection and speciation. *Annual Review of Ecology, Evolution, and Systematics* 38: 79–102. https://doi.org/10.1146/annurev.ecolsys.38.091206.095733
- Van Doorn, G.S., Edelaar, P. and Weissing, F.J. (2009) On the origin of species by natural and sexual selection. *Science* 326: 1704–1707. https://doi.org/10.1126/science.1181661

Sexual selection and ecology in the evolution of African lake cichlids: Wagner, C.E., Harmon, L.J. and Seehausen, O. (2012) Ecological opportunity and sexual selection together predict adaptive radiation. *Nature* 487: 366–370. https://doi.org/10.1038/nature11144

The spectacular evolution of African cichlids: Stiassny, M. and Meyer, A. (1999) Cichlids of the Rift Lakes. *Scientific American* 280: 64–69. www.jstor.org/stable/26058059

Percentages of endemic flowering plants of archipelagos: Lowry, P.P. (2009) Patterns of species richness, endemism, and diversification in oceanic island floras. In *Oceans and aquatic ecosystems. Vol. II. Encyclopedia of Life Support Systems*, pp. 201–220. UNESCO.

The endemicity of the Sierra Nevada flora: Blanca, G. (2002) *Flora amenazada y endémica de Sierra Nevada*. Granada: Consejería de Medio Ambiente.

Endemics of geologically old lakes: Rosenzweig, M.L. (1995) *Species Diversity in Space and Time*, p. 189.

Prevalence of allopatric speciation, and its significance in Australian mice: Barraclough, T.G. and Vogler, A.P. (2000) Detecting the geographical pattern of speciation from species-level phylogenies. *The American Naturalist* 155: 419–434. https://doi.org/10.1086/303332

Sympatric speciation in palms *Howea* and data on the floristic diversity of Lord Howe Island: Savolainen, V., Anstett, M., Lexer, C. *et al.* (2006) Sympatric speciation in palms on an oceanic island. *Nature* 441: 210–213. https://doi.org/10.1038/nature04566

Polyploid species: Rosenzweig, M.L. (1995) *Species Diversity in Space and Time*, pp. 48–49 (latitudinal gradient of polyploids in the world flora), 96–97 (speciation by polyploidisation), 338–341 (theory to explain the latitudinal gradient of polyploid plants).

Incidence of polyploidisation in plants: Wood, T.E., Takebayashi, N., Barker, M.S. *et al.* (2009) The frequency of polyploid speciation in vascular plants. *Proceedings of the National Academy of Sciences of the USA* 106: 13875–13879. https://doi.org/10.1073/pnas.0811575106

Hybrid speciation in sparrows: Hermansen, J. S., Sæther, S. A., Elgvin, T. O. *et al.* (2011). Hybrid speciation in sparrows I: phenotypic intermediacy, genetic admixture and barriers to gene flow. *Molecular Ecology* 20: 3812–3822. https://doi.org/10.1111/j.1365-294x.2011.05183.x

Chapter 3

The 'periodic tables of niches': Pianka, E.R. (2011) *Evolutionary Ecology*, 7th edition, pp. 290–291 (eBook).

Origin of the term 'ecological niche': Johnson, R.H. (1910) *Determinate Evolution in the Color-Pattern of the Lady-Beetles*. Washington DC: Carnegie Institution of Washington. https://doi.org/10.5962/bhl.title.30902

Darwin's version of the niches as 'positions in the economy of nature' appears in Chapter IV of *On the Origin of Species* (1859), under the heading 'Divergence of characters', third paragraph.

The Grinnellian niche: Grinnell, J. (1917) The niche relationships of the California thrasher. *The Auk* 34: 427–433. https://doi.org/10.2307/4072271

The Eltonian niche, and a vision of ecology that is surprisingly fresh and deep: Elton, C.S. (1927) *Animal Ecology*. New York: MacMillan.

The Hutchinsonian niche: Hutchinson, G.E. (1957) Concluding remarks. *Cold Spring Harbor Symposia on Quantitative Biology* 22: 415–427. http://symposium.cshlp.org/lookup/doi/10.1101/SQB.1957.022.01.039

Niche segregation in Daphne Major's finches: Grant, P.R. and Grant, B.R. (2006) Evolution of character displacement in Darwin's finches. *Science* 313: 224–226. https://doi.org/10.1126/science.1128374

Niche differentiation in salamanders of the genus *Plethodon*: Adams, D.C. and Rohlf, F.J. (2000) Ecological character displacement in *Plethodon*: biomechanical differences found from a geometric morphometric study. *Proceedings of the National Academy of Sciences of the USA* 97: 4106–4111. https://doi.org/10.1073/pnas.97.8.4106

First proposal of the term 'character displacement', although the roots of the idea go back to Darwin: Brown, W.L. and Wilson, E.O. (1956) Character displacement. *Systematic Zoology* 5: 49–64. https://doi.org/10.2307/2411924

A classic study of niche segregation in birds: MacArthur, R.H. (1958) Population ecology of some warblers of northeastern coniferous forests. *Ecology* 39: 599–619. https://doi.org/10.2307/1931600

Gause's experiments with competing microbes: Gause, G.F. (1934) *The Struggle for Existence*. Baltimore, MD: Williams & Wilkins.

Darwin's observation of competitive exclusion in herbs comes from Chapter III of *On the Origin of Species* (1859).

The niche of the barnacle *Chthamalus*: Connell, J.H. (1961) The influence of interspecific competition and other factors on the distribution of the barnacle *Chthamalus stellatus*. *Ecology* 42: 710–723. https://doi.org/10.2307/1933500

Predators can impede competitive exclusion: Paine, R.T. (1966) Food web complexity and species diversity. *The American Naturalist* 100: 65–75. https://doi.org/10.1086/282400

The intermediate disturbance hypothesis:
– Grime, J.P. (1973) Competitive exclusion in herbaceous vegetation. *Nature* 242: 344–347. https://doi.org/10.1038/242344a0
– Connell, J.H. (1978) Diversity in tropical rain forests and coral reefs. *Science* 199: 1302–1310. https://doi.org/10.1126/science.199.4335.1302

Testing the intermediate disturbance hypothesis:
– Sousa, W.P. (1979) Disturbance in marine intertidal boulder fields: the nonequilibrium maintenance of species diversity. *Ecology* 60: 1225–1239. https://doi.org/10.2307/1936969

- Wilkinson, D.M. (1999) The disturbing history of intermediate disturbance. *Oikos* 84: 145–147. https://doi.org/10.2307/3546874
- Rogers, C.S. (1993) Hurricanes and coral reefs: the intermediate disturbance hypothesis revisited. *Coral Reefs* 12: 127–137. https://doi.org/10.1007/BF00334471
- Bongers, F., Poorter, L., Hawthorne, W.D. and Shell, D. (2009) The intermediate disturbance hypothesis applies to tropical forests, but disturbance contributes little to tree diversity. *Ecology Letters* 12: 798–805. https://doi.org/10.1111/j.1461-0248.2009.01329.x

Fire and floristic diversity in the Mediterranean basin: Keeley, J.E., Bond, W.J., Brastock, R.A., Pausas, J.G. and Rundel, P.W. (2011) *Fire in Mediterranean Ecosystems: Ecology, Evolution and Management.* Cambridge: Cambridge University Press. https://doi.org/10.1017%2FCBO9781139033091

Forest bird communities in Hawaii: Moulton, M.P. and Pimm, S.L. (1987) Morphological assortment in introduced Hawaiian passerines. *Evolutionary Ecology* 1: 113–124. http://dx.doi.org/10.1007/BF02067395

Chapter 4

Bird species–area data for the plot: Preston, F.W. (1960) Time and space and the variation of species. *Ecology* 41: 611–627. https://doi.org/10.2307/1931793

A good review of the species–area curve: Rosenzweig, M.L. (1995) *Species Diversity in Space and Time*, Chapters I, VIII and IX.

Large-scale biodiversity: Gaston, K.J. and Blackburn, T.M. (2000) *Pattern and Process in Macroecology*, Chapter 2. Oxford: Blackwell. https://doi.org/10.1002/9780470999592

Philip Sclater's pioneering work delimiting the faunal regions of the world: Sclater, P.L. (1858) On the general geographical distribution of the members of the class Aves. *Journal of the Proceedings of the Linnean Society of London – Zoology* 2: 130–136. https://doi.org/10.1111/j.1096-3642.1858.tb02549.x

Wallace's regions: Wallace, A.R. (1876) *The Geographical Distribution of Animals*. New York: Harper & Brothers.

Diversity resonates at different scales; the more regional diversity, the more local diversity (usually): Gaston, K.J. and Blackburn, T.M. (2000) *Pattern and Process in Macroecology*, Chapter 2.

Chapter 5

The balance of species on islands was proposed by Munroe before MacArthur and Wilson:
- Munroe, E.G. (1948) The geographical distribution of butterflies in the West Indies. PhD thesis, Cornell University.
- Munroe, E.G. (1953) The size of island faunas. In: *Proceedings of the Seventh Pacific Science Congress of the Pacific Science Association*, Vol. IV. Zoology, pp. 52–53. Auckland: Whitcome & Tombs.
- Brown, J.H. and Lomolino, M.V. (1989) Independent discovery of the equilibrium theory of island biogeography. *Ecology* 70: 1954–1957. https://doi.org/10.2307/1938125

REFERENCES AND FURTHER READING

The book on the theory of island biogeography: MacArthur, R. and Wilson, E.O. (1967) *The Theory of Island Biogeography.* Princeton, NJ: Princeton University Press. https://doi.org/10.1515/9781400881376

Review of island biogeography 50 years after MacArthur and Wilson's book: Warren, B.H., Simberloff, D., Ricklefs, R.E. *et al.* (2015) Islands as model systems in ecology and evolution: prospects fifty years after MacArthur–Wilson. *Ecology Letters* 18: 200–217. https://doi.org/10.1111/ele.12398

The mangrove island experiment:
- Simberloff, D.S. and Wilson, E.O. (1969) Experimental zoogeography of islands: the colonization of empty islands. *Ecology* 50: 278–296. https://doi.org/10.2307/1934856
- Simberloff, D.S. (1974) Equilibrium theory of island biogeography and ecology. *Annual Review of Ecology and Systematics* 5: 161–182. https://doi.org/10.1146/annurev.es.05.110174.001113

The words of the exterminator of the mangrove islands, as told by MacArthur to his student: Rosenzweig, M.L. (1995) *Species Diversity in Space and Time*, p. 247.

The example of Krakatoa and the data on the distance effect in Oceania: MacArthur, R.H. and Wilson, E.O. (1963) An equilibrium theory of insular zoogeography. *Evolution* 17: 373–387. https://doi.org/10.1111/j.1558-5646.1963.tb03295.x

The rescue effect: Brown, J.H. and Kodrik-Brown, A. (1977) Turnover rates in insular biogeography: effect of immigration on extinction. *Ecology* 58: 445–449. https://doi.org/10.2307/1935620

Examples of how the variety of habitats promotes diversity on islands: Rosenzweig, M.L. (1995) *Species Diversity in Space and Time*, pp. 217–228.

Sparrows on the Canary Islands:
- Lack, D. (1969) The numbers of bird species on islands. *Bird Study* 16: 193–209. https://doi.org/10.1080/00063656909476244
- Bannerman, D.A. (1963) *A History of the Birds of the Canary Islands and of the Salvages.* Birds of the Atlantic Islands, Vol. 1. Edinburgh: Oliver & Boyd.

Ecological release of land birds in the West Indies: Cox, G.W. and Ricklefs, R.E. (1977) Species diversity and ecological release in Caribbean land bird faunas. *Oikos* 28: 113–122. https://doi.org/10.2307/3543330

Ecological release of anole lizards on Caribbean islands: Lister, B.C. (1976) The nature of niche expansion in West Indian *Anolis* lizards I: ecological consequences of reduced competition. *Evolution* 30: 659–676. https://doi.org/10.1111/j.1558-5646.1976.tb00947.x

The 'ultimate, all purpose bird': Lack, D. (1969) The numbers of bird species on islands. *Bird Study* 16: 193–209. https://doi.org/10.1080/00063656909476244

Proposal of the idea of biological resistance to invasion: Elton, C.S. (1958) *The Ecology of Invasions by Animals and Plants.* Chicago, IL: University of Chicago Press, reprint 2000. https://doi.org/10.1007/978-3-030-34721-5

Selected references on invasion resistance at different spatial scales:
- Stachowicz, J.J., Whitlatch, R.B. and Osman, R.W. (1999) Species diversity and invasion resistance in a marine ecosystem. *Science* 286: 1577–1579. https://doi.org/10.1126/science.286.5444.1577

- Levine, J.M. (2000) Species diversity and biological invasions: relating local process to community pattern. *Science* 288: 852–854. https://doi.org/10.1126/science.288.5467.852

The general dynamic model of island biogeography: Whittaker, R.J., Triantis, K.A. and Ladle, R.J. (2008) A general dynamic theory of oceanic island biogeography. *Journal of Biogeography* 35: 977–994. http://dx.doi.org/10.1111/j.1365-2699.2008.01892.x

Chapter 6

The quotations from Darwin come from the first paragraph of Chapter XI of *On the Origin of Species* (1859), under the heading 'On extinction'.

Information on the 'Akiapola'au: BirdLife International. Species factsheet: *Hemignathus wilsoni*. https://datazone.birdlife.org/species/factsheet/akiapolaau-hemignathus-wilsoni

The 'I'iwi and its conservation status. https://datazone.birdlife.org/species/factsheet/iiwi-drepanis-coccinea

'Speciation rebound' after mass extinctions:

- Sepkoski, J.J. (1998) Rates of speciation in the fossil record. *Philosophical Transactions of the Royal Society* B353: 315–326. https://doi.org/10.1098/rstb.1998.0212
- Erwin, D.H. (1998) The end and the beginning: recoveries from mass extinctions. *Trends in Ecology and Evolution* 13: 344–349. https://doi.org/10.1016/S0169-5347(98)01436-0
- Kirchner J.W. and Weil, A. (2000) Delayed biological recovery from extinctions throughout the fossil record. *Nature* 404: 177–180. https://doi.org/10.1038/35004564

The evolution of mammals after the Cretaceous–Palaeogene extinction:

- Alroy, J. (1999) The fossil record of North American mammals: evidence for a Paleocene evolutionary radiation. *Systematic Biology* 48: 107–118. https://doi.org/10.1080/106351599260472
- Meredith, R.W., Janečka, J.E., Gatesy, J. *et al.* (2011) Impacts of the Cretaceous terrestrial revolution and KPg extinction on mammal diversification. *Science* 334: 521–524. https://doi.org/10.1126/science.1211028
- A much-cited criticism of the evolutionary explosion of mammals after that extinction based on molecular data: Bininda-Emonds, O.R., Cardillo, M., Jones, K.E. *et al.* (2007) The delayed rise of present-day mammals. *Nature* 446: 507–512. https://doi.org/10.1038/nature05634

The risk of extinction increases with the diversity of trilobites at the Murero site: López-Villalta, J.S. (2016) Self-regulation of trilobite diversity in Murero (middle Cambrian, Spain) due to compensatory extinction. *Geologica Acta* 14: 71–78. https://doi.org/10.1344/GeologicaActa2016.14.1.6

Relationships between diversity, speciation and extinction in marine invertebrates in the fossil record: Foote, M. (2010) The geologic history of diversity. In Bell, M., Futuyma, D., Eanes, W. and Levinton, J. (eds.), *Evolution Since Darwin: the First 150 Years*, pp. 479–510. Sunderland, MA: Sinauer Associates.

The evolution of Cenozoic terrestrial mammal families: Quental, T.B. and Marshall, C.R. (2013) How the Red Queen drives terrestrial mammals to extinction. *Science* 341: 290–292. https://doi.org/10.1126/science.1239431

Marine animal genera become more enduring: Foote, M. (2010) The geologic history of diversity. In Bell M., Futuyma, D., Eanes, W. and Levinton, J. (eds.), *Evolution Since Darwin: the First 150 Years*, pp. 479–510.

Diversity of fossil marine invertebrates, and specialisation: Nürnberg, S. and Aberhan, M. (2015) Interdependence of specialization and biodiversity in Phanerozoic marine invertebrates. *Nature Communications* 6: 6602. https://doi.org/10.1038/ncomms7602

In the fossil communities from Mistaken Point, the diversity was already typical of its modern analogues: Clapham, M.E., Narbonne, G.M. and Gehling, J.G. (2003) Paleoecology of the oldest known animal communities: Ediacaran assemblages at Mistaken Point, Newfoundland. *Paleobiology* 29: 527–544. https://doi.org/10.1666/0094-8373(2003)029%3C0527:POTOKA%3E2.0.CO;2

Spores of the oldest plants: Steemans, P., Hérissé, A.L., Melvin, J. et al. (2015) Origin and radiation of the earliest vascular land plants. *Science* 324: 353. https://doi.org/10.1126/science.1169659

Data in Figure 6.5: Alroy, J., Aberhan, M., Bottjer, D.J. et al. (2008) Phanerozoic trends in the global diversity of marine invertebrates. *Science* 321: 97–100. https://doi.org/10.1126/science.1156963

The first fossil diversity curve: Phillips, J. (1860) *Life on the Earth: its Origin and Succession.* London: MacMillan. https://doi.org/10.5962/BHL.TITLE.22153

The ecological niches of marine fauna over time: Bambach, R.K., Bush, A.M. and Erwin, D.H. (2007) Autecology and the filling of ecospace: key metazoan radiations. *Palaeontology* 50: 1–22. https://doi.org/10.1111/j.1475-4983.2006.00611.x

Diversification by addition of new niches rather than by accumulation of species into existing niches: Bambach, R.K. (1983) Ecospace utilization and guilds in marine communities through the Phanerozoic. In Tevesz, M.J.S and McCall, P.L. *Biotic Interactions in Recent and Fossil Benthic Communities*, pp. 719–746. Boston, MA: Springer. https://doi.org/10.1007/978-1-4757-0740-3_15

The diversity of marine invertebrate communities has undoubtedly increased throughout the history of life: Bush, A.M. and Bambach, R.K. (2004) Did alpha diversity increase during the Phanerozoic? Lifting the veils of taphonomic, latitudinal, and environmental biases. *Journal of Geology* 112: 625–642. https://doi.org/10.1086/424576

Enter the Court Jester: Barnosky, A.D. (2001) Distinguishing the effects of the Red Queen and Court Jester on Miocene mammal evolution in the Northern Rocky Mountains. *Journal of Vertebrate Paleontology* 21: 172–185. http://dx.doi.org/10.1671/0272-4634(2001)021[0172:DTEOTR]2.0.CO;2

The Red Queen and the Court Jester: Benton, M.J. (2009) The Red Queen and the Court Jester: species diversity and the role of biotic and abiotic factors through time. *Science* 323: 728–732. https://doi.org/10.1126/science.1157719

Chapter 7

Humboldt and the tropical diversity gradient: Hawkins, B.A. (2001) Ecology's oldest pattern? *Trends in Ecology and Evolution* 16: 470. https://doi.org/10.1016/s0160-9327(00)01369-7

The latitudinal gradient in the history of the Earth: Mannion, P.D., Upchurch, P., Benson, R.B.J. and Goswami, A. (2014) The latitudinal biodiversity gradient through deep time. *Trends in Ecology and Evolution* 29: 42–50. https://doi.org/10.1016/j.tree.2013.09.012

An excellent review of the possible reasons for the latitudinal gradient. Willig, M.R., Kaufman, D.M. and Stevens, R.D. (2003) Latitudinal gradients of biodiversity: pattern, process, scale, and synthesis. *Annual Review of Ecology, Evolution, and Systematics* 34: 273–309. https://doi.org/10.1146/annurev.ecolsys.34.012103.144032

The latitudinal gradient and evolution: Mittelbach, G.G., Schemske, D.W. Cornell, H.V. et al. (2007) Evolution and the latitudinal diversity gradient: speciation, extinction, and biogeography. *Ecology Letters* 10: 315–331. https://doi.org/10.1111/j.1461-0248.2007.01020.x

Speciation, extinction and the tropical gradient in mammals: Rolland, J., Condamine, F.L., Jiguet, F. and Morlon, H. (2014) Faster speciation and reduced extinction in the tropics contribute to the mammalian latitudinal diversity gradient. *PlOS Biology* 12: e1001775. https://doi.org/10.1371/journal.pbio.1001775

The tropics, cradle and museum of species: Stebbins, G.L. (1974) *Flowering Plants: Evolution Above the Species Level.* Cambridge, MA: Harvard University Press.

The Andes uplift and the diversification of Amazonian fauna and flora: Hoorn, C., Wesselingh, F.P., ter Steege, H. et al. (2010) Amazonia through time: Andean uplift, climate change, landscape evolution, and biodiversity. *Science* 330: 927–931. https://doi.org/10.1126/science.1194585

Most marine animal lineages originated in the tropics: Jablonski, D. (1993) The tropics as a source of evolutionary novelty through geological time. *Nature* 364: 142–144. https://doi.org/10.1038/364142a0

Extinction of subtropical forest species in the Mediterranean zone: Mai, D.H. (1989) Development and regional differentiation of the European vegetation during the Tertiary. *Plant Systematics and Evolution* 162: 79–91. https://doi.org/10.1007/978-3-7091-3972-1_4

Why some habitats are more diverse than others: Terborgh, J. (1973) On the notion of favorableness in plant ecology. *The American Naturalist* 107: 481–501. https://doi.org/10.1086/282852

The rules of macroevolution:

– Stanley, S.M. (1979) *Macroevolution: Pattern and Process.* Baltimore, MD: Johns Hopkins University Press. 1998 edition.

– Stanley, S.M. (1990) The general correlation between rate of speciation and rate of extinction: fortuitous causal linkages. In Ross, R.M. and Allmon, W.D. (eds.), *Causes of Evolution: a Paleontological Perspective*, pp. 103–127. Chicago, IL: University of Chicago Press.

Data in Table 7.1:

– Adl, S.M., Simpson, A.G.B., Lane, C.E. et al. (2012) The revised classification of eukaryotes. *Journal of Eukaryotic Microbiology* 59: 429–493. https://doi.org/10.1111/j.1550-7408.2012.00644.x

– Derelle, R., López-García, P., Timpano, H. and Moreira, D. (2016) A phylogenomic framework to study the diversity and evolution of stramenopiles (=heterokonts). *Molecular Biology and Evolution* 33: 2890–2898. https://doi.org/10.1093/molbev/msw168

- Dunn, C.W., Giribet, G., Edgecombe, G.D. and Hejnol, A. (2014) Animal phylogeny and its evolutionary implications. *Annual Review of Ecology and Systematics* 45: 371–395. http://dx.doi.org/10.1146/annurev-ecolsys-120213-091627
- Pryer, K.M., Schneider, H., Smith, A.R. et al. (2001) Horsetails and ferns are a monophyletic group and the closest living relatives of seed plants. *Nature* 409: 618–622. https://doi.org/10.1038/35054555
- Bell, G. and Mooers, A.O. (1997) Size and complexity among multicellular organisms. *Biological Journal of the Linnean Society* 60: 345–363. https://doi.org/10.1006/bijl.1996.0108
- Valentine, J.W. (2003) Cell types, numbers, and body plan complexity. In Hall, B.K. and Olson, W.M. (eds.), *Keywords and Concepts in Evolutionary Developmental Biology*, pp. 35–43. Cambridge, MA: Harvard University Press. https://doi.org/10.4159/9780674273320-008
- Margulis, L. and Chapman, M.J. (2009) *Kingdoms & Domains: an Illustrated Guide to the Phyla of Life on Earth.* Academic Press. Corrected edition, 2010.

Factors promoting diversification:
- Animals: McPeek, M.A. and Brown, J.M. (2007) Clade age and not diversification rate explains species richness among animal taxa. *The American Naturalist* 169: 97–106. https://doi.org/10.1086/512135
- Plants: Givnish, T.J. (2010) Ecology of plant speciation. *Taxon* 59: 1326–1366. https://doi.org/10.1002/tax.595003

Chapter 8

Data on endangered and extinct species: IUCN Red List version 2024-2: Table 1a and Table 3. www.iucnredlist.org/resources/summary-statistics

Global biodiversity hotspots: Mittermeier, R., Turner, W.R., Larsen, F.W., Brooks, T.M. and Gascon, C. (2011) Global biodiversity conservation: the critical role of hotspots. In Zachos, F.E. and Habel, J.C. (eds.), *Biodiversity Hotspots*, pp. 3–22. Berlin: Springer. http://dx.doi.org/10.1007/978-3-642-20992-5_1

Numbers of known and unknown species: Costello, M.J., May, R.M. and Stork, N.E. (2013) Can we name Earth's species before they go extinct? *Science* 339: 413–416. https://doi.org/10.1126/science.1230318

Human impact on biodiversity: Gaston, K.J. and Spicer, J.I. (2013) *Biodiversity: an Introduction*, 2nd edition, Chapter 5.

Data on habitat loss: Pereira, M., Navarro, L.M. and Martins, I.S. (2012) Global biodiversity change: the good, the bad, and the unknown. *Annual Review of Environment and Resources* 37: 25–50. https://doi.org/10.1146/annurev-environ-042911-093511

History of the Thylacine: *The Thylacine Museum.* www.naturalworlds.org/thylacine.

The mosquito and the birds of Hawaii: Warner, R.E. (1968) The role of introduced diseases in the extinction of the endemic Hawaiian avifauna. *The Condor* 70: 101–120. https://doi.org/10.2307/1365954

The story of *Boiga irregularis* in Guam: Pimm S.L. (1987) The snake that ate Guam. *Trends in Ecology and Evolution* 2: 293–295. http://dx.doi.org/10.1016/0169-5347(87)90080-2

Review of the biological invasion of Guam: Fritts, T.H. and Rodda, G.H. (1998) The role of introduced species in the degradation of island ecosystems: a case history

of Guam. *Annual Review of Ecology and Systematics* 29: 113–140. https://doi.org/10.1146/annurev.ecolsys.29.1.113
The top 100 invasive species, according to the IUCN: Global Invasive Species Database (2025). www.iucngisd.org/gisd/100_worst.php
The book that was the spark for our interest in biological invasions: Elton C.S. (1958) *The Ecology of Invasions by Animals and Plants*. Chicago, IL: University of Chicago Press, reprint 2000. https://doi.org/10.1007/978-3-030-34721-5
The '10% rules' of biological invasions: Williamson, M., and Fitter, A. (1996) The varying success of invaders. *Ecology* 77: 1661–1666. https://doi.org/10.2307/2265769
The rise and fall of the Passenger Pigeon:
- Rosenzweig M.L. (1995) *Species Diversity in Space and Time*, pp. 122–123 and 125–126.
- Schorger, A.W. (1955) *The Passenger Pigeon: its Natural History and Extinction*. Madison, WI: University of Wisconsin Press.
- Avery, M. (2014) *A Message from Martha: the Extinction of the Passenger Pigeon and its Relevance Today*. London: Bloomsbury.
IPCC Global Synthesis Report on Climate Change: IPCC (2014) *Climate Change 2014: Synthesis Report*. Geneva: IPCC. www.ipcc.ch
The threat of climate change according to IUCN data: Foden, W., Mace, G., Vié, J.-C. et al. (2008) Species susceptibility to climate change impacts. In Vié, J.-C., Hilton-Taylor, C. and Stuart, S.N. (eds), *The 2008 Review of The IUCN Red List of Threatened Species*. Gland: IUCN. https://doi.org/10.2305/IUCN.CH.2009.17.en
The decline of *Lottia alveus*: Carlton, J.T., Vermeij, G.J., Lindberg, D.R., Carlton, D.A. and Dubley, E.C. (1991) The first historical extinction of a marine invertebrate in an ocean basin: the demise of the eelgrass limpet *Lottia alveus*. *Biological Bulletin* 180: 72–80. https://doi.org/10.2307/1542430
Co-extinction of species: Koh, L.P., Dunn, R.R., Sodhi, N.J. et al. (2004) Species coextinctions and the biodiversity crisis. *Science* 305: 1632–1634. https://doi.org/10.1126/science.1101101
The text of the Convention on Biological Diversity: Secretariat of the Convention on Biological Diversity (2011) *Convention on Biological Diversity: text and annexes*. United Nations Environment Programme. www.cbd.int
Medicines from biodiversity: Wilson, E.O. (2001) *The Diversity of Life*, Chapter 13. London: Penguin.
Biodiversity conservation and societal change: Gaston, K.J. and Spicer, J.I. (2013) *Biodiversity: an Introduction*, 2nd edition, Chapters 5 and 6.

Glossary

Alpha, beta and gamma diversity, three unnecessarily complex-sounding terms proposed in a great article: Whittaker, R.H. (1972) Evolution and measurement of species diversity. *Taxon* 21: 213–251. https://doi.org/10.2307/1218190
Macroecology:
- Brown, J.H. (1995) *Macroecology*. Chicago, IL: University of Chicago Press. Spanish edition, 2004.
- Gaston, K.J. and Blackburn, T.M. (2000) *Pattern and Process in Macroecology*.

Metabolic theory of ecology: Sibly, R.M., Brown, J.H. and Kodric-Brown, A. (eds.) (2012) *Metabolic Ecology: a Scaling Approach*. Chichester: Wiley-Blackwell. https://doi.org/10.1002/9781119968535

Mid-domain effect: Colwell, R.K. and Lees, D.C. (2000) The mid-domain effect: geometric constraints on the geography of species richness. *Trends in Ecology and Evolution* 15: 70–76. https://doi.org/10.1016/S0169-5347%2899%2901767-X

Paradox of the plankton
– Hutchinson, G.E. (1961) The paradox of the plankton. *The American Naturalist* 95: 137–145. https://doi.org/10.1086/282171
– Mitchell, J.G., Yamazaki, H., Seuront, L., Wolk, F. and Li, H. (2008) Phytoplankton patch patterns: seascape anatomy in a turbulent ocean. *Journal of Marine Systems* 69: 247–253. https://doi.org/10.1016/J.JMARSYS.2006.01.019

Peninsula effect: Gaston, K.J. and Spicer, J.I. (2013) *Biodiversity: an Introduction*, 2nd edition.

Punctuated equilibrium:
– Gould, S.J. and Eldredge, N. (1972) Punctuated equilibria: an alternative to phyletic gradualism. In Schopf, T.J.M (ed.), *Models in Paleobiology*, pp. 82–115. San Francisco, CA: Freeman.
– Stanley, S.M. (1979) *Macroevolution: Pattern and Process*. Baltimore, MD: Johns Hopkins University Press. 1998 edition.

Species–area curve: Rosenzweig, M.L. (1995) *Species Diversity in Space and Time*, Chapters I, VIII and IX.

Species selection: Jablonski, D. (2008) Species selection: theory and data. *Annual Review of Ecology, Evolution, and Systematics* 39: 501–524. https://doi.org/10.1146/ANNUREV.ECOLSYS.39.110707.173510

Unified neutral theory of biodiversity and biogeography:
– Hubbell, S.P. (2001) *The Unified Neutral Theory of Biodiversity and Biogeography*. Princeton, NJ: Princeton University Press.
– McGill, B.J. (2003) A test of the unified neutral theory of biodiversity. *Nature* 422: 881–885. https://doi.org/10.1038/nature01583
– Dornelas, M., Connolly, S. and Hughes, T. (2006) Coral reef diversity refutes the neutral theory of biodiversity. *Nature* 440: 80–82. https://doi.org/10.1038/nature04534

Illustration credits

The Creative Commons images used in this book have not been digitally modified. The use of Creative Commons images requires that the author is credited and that his or her rights are respected under the terms of the specific licence for each file.

Links to the legal text of the Creative Commons licences of the images reproduced in this book:

CC0 1.0 Universal (dedication to the public domain): https://creativecommons.org/publicdomain/zero/1.0/legalcode

CC BY 2.0 Creative Commons – Attribution 2.0 Generic: https://creativecommons.org/licenses/by/2.0/legalcode

CC BY-SA 2.0 Creative Commons Attribution-ShareAlike 2.0 Generic: https://creativecommons.org/licenses/by-sa/2.0/legalcode

CC BY 3.0 Creative Commons – Attribution 3.0 Unported: https://creativecommons.org/licenses/by/3.0/legalcode

CC BY-SA 3.0 Creative Commons – Attribution-ShareAlike 3.0 Unported: https://creativecommons.org/licenses/by-sa/3.0/legalcode

CC BY-SA 4.0 Creative Commons – Attribution-ShareAlike 4.0 International: https://creativecommons.org/licenses/by-sa/4.0/legalcode

***

Figure 1.1 Image in the public domain, since the artist died more than 100 years ago, in 1841.

Figure 1.2 The Image of the light form is the photograph 'Biston betularia (2938017418).jpg', by Donald Hobern, showing a female from Hellerup, Denmark, licensed under CC BY 2.0: https://upload.wikimedia.org/wikipedia/commons/d/df/Biston_betularia_%282938017418%29.jpg. The image of the dark form is 'Biston betularia(js) o1 Lodz(Poland).jpg', obtained in Lodz, Poland, by Jerzy Strzelecki and licensed under CC BY 3.0: https://upload.wikimedia.org/wikipedia/commons/d/d5/Biston_betularia%28js%2901_Lodz%28Poland%29.jpg

Figure 1.3 The salmon drawing is in the public domain as an illustration created by the US government for publication, appearing on p. 4 of pamphlet 1996-792-501: *Lake Washington Ship Canal Fish Ladder*. The Guppies are the photograph 'Guppy pho 0048' by Per Harald Olsen, licensed under CC BY 3.0: https://commons.m.wikimedia.org/wiki/File:Guppy_pho_0048.jpg

Figure 1.4 Image in the public domain, since the artist died in 1915.

ILLUSTRATION CREDITS

Figure 2.1 Image in the public domain, since the artist died in 1881.
Figure 2.2 Image in the public domain, since the artist died in 1917.
Figure 2.3 Image in the public domain, since the artist died in 1917.
Figure 2.4 Reproduced with permission of the photographer, Ben Twist.
Figure 2.5 Reproduced with permission of the photographer, Ian Hutton.
Figure 3.1 Author's own workings.
Figure 3.2 Reproduced with permission of the artist, Deborah Kaspari.
Figure 3.3 Author's own workings.
Figure 3.4 The photograph is the image 'Borneo rainforest.jpg', by Dukeabruzzi, showing a rainforest in Kinabalu Park, licensed under CC BY-SA 4.0: https://commons.wikimedia.org/wiki/File:Borneo_rainforest.jpg
Figure 4.1 Author's own workings.
Figure 4.2 The map of the faunal regions comes from an image in the public domain in the USA because it was published before 1923. It comes from p. 312 of vol. 6 of *The New International Encyclopaedia*, 1905, based on data from Philip Sclater and Alfred R. Wallace. Photographs of marsupials are reproduced with the consent of the photographer, Ben Twist.
Figure 5.1 The top two images are in the public domain, obtained by NASA (National Aeronautic and Space Administration, USA). The Hawaiian landscape is the image 'Kauai, forest.jpg', a forest near Hanalei Bay, Kauai, Hawaii, by Lukas, licensed under CC BY 2.0: https://commons.m.wikimedia.org/wiki/File:Kauai,_forest.jpg. The waterfall looming between the hillsides is the one featured in the movie *Jurassic Park* (!). The landscape of the Canary Islands is an image bequeathed by its author, Garavitotfe, to the public domain under a CC0 1.0 licence.
Figure 5.2 Author's own workings.
Figure 5.3 The photograph of Surtsey is a public domain image from NOAA (National Oceanic and Atmospheric Administration, USA). The Atlantic Puffin is 'Papageitaucher Fratercula arctica.jpg', by Richard Bartz, taken at Látrabjarg, Iceland, and licensed under CC BY-SA 3.0: https://commons.m.wikimedia.org/wiki/File:Papageitaucher_Fratercula_arctica.jpg
Figure 5.4 The Bananaquits are from the photograph 'Two Bananquits (*Coereba flaveola*) on a branch – Campo Limpo Paulista Sao Paulo, Brazil' by Leon-bojarczuk – Flickr page, licensed under CC BY-SA 2.0: https://upload.wikimedia.org/wikipedia/commons/b/b5/Bananaquits.jpg. The Jamaican Tody is the image 'Jamaican tody (Todus todus), Jamaica', by Charlesjsharp, licensed under CC BY-SA 4.0: https://upload.wikimedia.org/wikipedia/commons/7/75/Jamaican_tody_%28Todus_todus%29.jpg. The Cuban Green Anole is the photograph 'Anolis porcatus on leaves in Cuba', by Thomas H Brown, licensed under CC BY 2.0: https://upload.wikimedia.org/wikipedia/commons/e/e4/Anolis_porcatus_on_leaves_in_Cuba.jpg. The Tocororo trogon is the image 'A Cuban trogon in Camagüey, Camagüey province, Cuba', by Laura Gooch, licensed under CC BY-SA 2.0: https://commons.wikimedia.org/wiki/File:Priotelus_temnurus_-Camaguey,_Camaguey_Province,_Cuba-8.jpg
Figure 6.1 Author's own workings.
Figure 6.2 Image of the Marine Iguana reproduced with permission of the photographer, Ben Twist. The Kea is from the photograph 'Kea on rock while snowing' by Alan Liefting, released by its author to the public domain. The image of the 'I'iwi is a public domain photograph from the US Geological Survey. The Kākāpō is called Sirocco, lives on Maud Island, and its photo, 'Kakapo Sirocco 1', was obtained by Chris Birmingham,

from the New Zealand Department of Conservation and licensed under CC BY-SA 2.0: https://commons.wikimedia.org/wiki/File:Kakapo_Sirocco_1.jpg

Figure 6.3 Images from a series of stickers from 1916–1920(?), which are in the public domain in most countries because the author died more than 70 years ago, in 1935.

Figure 6.4 Reproduced with permission of the photographer, Alex Liu.

Figure 6.5 Drawings from Sowerby, J. (1812) *The Mineral Conchology of Great Britain*, in the public domain due to the death of its author more than 100 years ago, in 1822. Plots drawn from the data cited in the references.

Figure 7.1 Image in the public domain due to the death of the artist more than 100 years ago, in 1900.

Figure 7.2 Author's own workings.

Figure 8.1 The map 'Biodiversity hotspots.svg' is attributed to Crates and is licensed under CC BY-SA 2.0: https://commons.wikimedia.org/wiki/File:Biodiversity_Hotspots.svg

Figure 8.2 All images are in the public domain. The photograph of a male and female Thylacine at the Washington DC Zoo was obtained by Baker and E.J. Keller around 1904. The photograph of the Passenger Pigeon was taken by Hayashi and Toda in 1920. The Great Auks are from a classic illustration by John James Audubon, who died more than 100 years ago, in 1851. The painting of the Dodo is by Roelant Savery (1576–1639).

Figure 8.3 The photograph of the snake is 'Brown tree snake (Boiga irregularis) (8387575202).jpg', Bogani Nani Wartabone National Park, taken by Pavel Kirillov and licensed under CC BY-SA 2.0: https://upload.wikimedia.org/wikipedia/commons/f/ff/Brown_tree_snake_(Boiga_irregularis)_(8387575202).jpg. The Common Water Hyacinth is 'Eichhornia crassipes 003.jpg', by H. Zell, licensed under CC BY-SA 3.0: https://upload.wikimedia.org/wikipedia/commons/8/8d/Eichhornia_crassipes_003.JPG. The image of the European Rabbits, whose author is unknown, has the number 11145789 in the National Archives of Australia and is in the public domain due to copyright expiry. The Northern Pacific Sea Star is a photograph released by the photographer (probably Lycoo) to the public domain.

Figure 8.4 The Polar Bear photo is 'Polar Bear ANWR 1.jpg', taken by Alan D. Wilson at the Arctic National Wildlife Refuge, Alaska, and licensed under CC BY-SA 3.0: https://upload.wikimedia.org/wikipedia/commons/4/45/Polar_Bear_ANWR_1.jpg. The corals are 'Coral Outcrop Flynn Reef.jpg', photographed by Toby Hudson near Cairns, Queensland, Australia, licensed under CC BY-SA 3.0: https://upload.wikimedia.org/wikipedia/commons/2/2e/Coral_Outcrop_Flynn_Reef.jpg. The image of the Golden Toad, taken by Charles H. Smith, is in the public domain as it was obtained by a US government agency (US Fish and Wildlife Service) as part of its public work. The photograph of the Iberian Lynx is from the Iberian Lynx Ex-Situ Conservation Programme (www.lynxexsitu.es), which authorises its use for any purpose provided the programme is named as the copyright holder.

Figure 8.5 The photo of the Grey Whale is 'Escrichtius robustus 01.jpg', by Merrill Gosho (NOAA), being in the public domain as it was obtained during the work of a US public government agency. The Black-footed Ferret is 'Black-footed Ferret (5244705122).jpg' by J. Michael Lockhart (USFWS), being in the public domain for the same reason. The image of Przewalski's Horse was taken in Hustain Nuruu National Park (Mongolia) and is 'Przewalski mongolie.jpg', by Bouette, licensed under CC BY-SA 3.0: https://commons.wikimedia.org/wiki/File:Przewalski_mongolie.jpg

Index

References to figures appear in *italic* type; those in **bold** type refer to tables.

Abrojo *Tribulus cistoides* 24, *24*
abundance of species 34, 38, 48, 68
African Great Lakes 16, 17
African megafauna 68
ageing 52
Agenda 21 programmes 76–7
'Akiapola'au *Hemignathus wilsoni* 48
alien species 44, 68, 70–2
allopatric speciation *17*, 19
ammonites *56*
amphibians 66–8, 73–4, **73**
anole lizards *43*, 44
Antarctic mountains 60
Archegosaurus 52
Arctic Sea Rocket *Cakile arctica 40*
Arthropoda (arthropods) **64**
Atlantic Puffin *Fratercula arctica 40*
Audubon, John James 72
Australia *17*, *35*, *36*, 70
Austrominius modestus 28
autopolyploids 20
Avalon biota 53

Bambach, Richard 55
Bananaquits *Coereba flaveola 43*
barnacles *28*, 29, 30
Barnosky, Anthony 57
Barylambda 51
Bay-breasted Warbler *Setophaga castanea 26*
Beadlet Anemones *Actinia equina 28*
biodiversity ix

and competitive exclusion 29
complexity and diversity 61
defining 1–2
disturbances (storms) 29–31, *31*
dynamic equilibrium 40
evolutionary balance *47*
habitats 60
intrinsic and functional value 74–5
threats to 66–9, *67*
'biological explosions' 72
biological species concept 12–13
birds 32, *33*, 34, 43–4, 66–8, **73**
birds of paradise *8*, *9*
Black-footed Ferret *Mustela nigripes 76*
Black Rat *Rattus rattus* 70
Black-throated Green Warbler *Setophaga virens 26*
Blackburnian Warbler *Setophaga fusca 26*
Blue-breasted Fairywren *Malurus pulcherrimus 17*
body shapes and sizes 32, 61
Borneo *31*
Brachiopoda (brachiopods) *56*, **64**
British Columbia, Canada 13
Brook Lamprey *Lampetra planeri* 14, *15*
Brown Tree Snake *Boiga irregularis* 70, *71*
Bumpus, Hermon 5

Cactus Finches *Geospiza scandens 2*, *3*
California 30
Cambrian period 51, 55
Cañadas del Teide, Tenerife *37*
Canary Islands *37*, 42–3
cancer treatments 75
Cape May Warbler *Setophaga tigrina 26*
Caribbean 43–4
Carroll, Lewis 53
character displacements 23–5
Charniodiscus 54
Chthamalus barnacles 27–9, *28*
Church, Frederic Edwin 59
cichlid fish 16, 17
Cíes Islands, Galicia, Spain *28*
climate change 72–4, **73**
clubmoss seedless plants 55
Coho Salmon *Oncorhynchus kisutch* 7–9, *7*
colonisation 40, *40*, 44, 55
colour blindness (achromatopsia) 5
Common Water Hyacinth *Eichhornia crassipes* 71
competition 43–4, *43*, 48
competitive exclusion 27, 29
complex organisms 61–2, *63*, **64**
Connell, Joseph 27, 30
conservation 76–7, *76*
Convention on Biological Diversity (UN) 1–2, 69–70, 74, 75, 77
corals and coral reefs 30, *56*, 59, *73*, **73**

Cork Oak *Quercus suber* 31
Court Jester hypothesis 57
Cox, George 43
'cradle effect' 59
Craniata (vertebrates and hagfishes) **64**
Creationism 6–7
Cretaceous–Palaeogene (K–Pg) extinction event 50
Cuban Green Anole *Anolis porcatus* 43

Daphne Major island 23–4, *24*
Darwin, Charles 2–4, 9, 10, 12, 14–15, 22, 27, 46, 53, 57
Darwin's finches *11*, 23–5, *24*
diet 61
Dingo 70
dinosaurs 50–1
direct persecution 68, 72
Discovery Bay, Jamaica 30
diseases 69
dispersal ability 61
'distance effect' 41–2
disturbances 29–30
diversity
  abundance 34
  complexity 61–3, **64**
  economic value 75
  extinction rates 38, *39*, 51
  habitats 60
  islands 38–42
  and new species 48
  resistance to invasion 44–5
  and speciation 50
  tropics 58–61
diversity curve *56*, 57
diversity equilibrium 46
Dodo *Raphus cucullatus* 68
domestic cat 70

echinoderms **64**
ecological niches 22–3
  community organisation 31–2
  competitors 43–4, *43*, 48
  marsupials *35*
  Mediterranean shrubland **22**
  shallow-sea fauna 55
ecological release 42–4

ecological speciation 12, 59
  see also speciation
The Ecology of Invasions by Animals and Plants (Elton) 71–2
Ediacaran period 53
Eelgrass *Zostera marina* 74
elephants 4
Elton, Charles 23, 71–2
endemic species 16–17
Eohippus 52
the equator 58, 60
Eupithecia palikea 48
European Green Crab *Carcinus maenas* 28
European Rabbits *Oryctolagus cuniculus* 71
European River Lamprey *Lampetra fluviatilis* 14, *15*
evolution 5, 12–13, 61–2
  see also natural selection
evolutionary equilibrium 51
evolutionary radiation 50, 54–5, *63*
expansive equilibrium 57
'extinction cascade' 74
extinction rates 41, 62, *63*, 66
extinction risk 38, *39*, 47, 48
extinctions 32, *47*, 68–9, *68*, 72–4
extreme habitats 61

famines 4, 24
faunal regions *35*, 36
Fiji 41
finches (Fringillidae) *11*, 23–5, *24*, 48–50
Fiordland Penguin *Eudyptes pachyrhynchus* 50
fires 31
Fitter, Alastair 72
Florida Keys 39–40
food 29, 75
Foote, Michael 51
fossils 51, *56*, *63*
founder effect 5
Fractofusus 54
fragmentation 69
freshwater fish 66
fungi **64**

Galápagos Islands *11*, 23, 25
game of species 1–2, 10, 22, 36, *37*–8, 63–5

Gause, Georgii Frantsevich 25–7
genetic drift 5
genetic homogenisation 16
Genovesa, Galápagos archipelago 2
geographic isolation 16–19
The Geographical Distribution of Animals (Wallace) 36
Ghana 30–1
Giant African Land Snail *Achatina fulica* 70–1
global warming 72–3
Golden Toad *Incilius periglenes* 73
Gough Finch *Rowettia goughensis* 44
Gough Island, South Atlantic 44
Gould, John *11*
granivorous (seed eating) birds 24, *24*
Grant, Peter and Rosemary 2, 25
Great Auk *Pinguinus impennis* 68
Great Cumbrae, Firth of Clyde, Scotland 27
Grey Whale *Eschrichtius robustus* 76
Grinnell, Joseph 22–3
Guam, Pacific island 70
Guppy *Poecilia reticulata* 7, 9–10
gymnosperms 66

habitat destruction 68, 69
habitats 55, 60–1
Hawaii 16, 32, *37*, 41, 70, 72
Hemichordata (hemichordates) **64**
herbivorous mammals 50
Holm Oak *Quercus ilex/rotundifolia* 31
honeycreepers (finches) 48
horsetail seedless plants 55, **64**
House Sparrow *Passer domesticus* 5, 21
Howea belmoreana 18, 19
Howea forsteriana 18, 19
human enterprise 76
humans *Homo sapiens* 1, 68–9
Humboldt, Alexander von 58
hunting 72

INDEX

Hurricane Allen 30
Hutchinson, George Evelyn 23
hybridogenesis 21

Iberian Lynx *Lynx pardinus* 73
'I'iwi *Drepanis coccinea* 48, *49*
immigration and extinction equilibrium 38, *39*
immigration rates 38, *39*, 41–2
industrialisation 6
insects 55, 61–2
instant species 19–21
Intergovernmental Panel on Climate Change (IPCC) 73
intermediate disturbance hypothesis 30–1
intermediate islands 41–2
International Union for Conservation of Nature (IUCN) 66, 68, 73
interspecies competition 29
invasive species 44, 68, 70–2
island biogeography theory 41–2
island habitats 16–17
island species 70
islands 17, 37–45, 48–51, 66
Italian Sparrow *Passer italiae* 21

Jamaican Tody *Todus todus* 43
Johnson, Roswell Hill 22
jungles 60

Kai (Kei) Islands 41
Kākāpō *Strigops habroptilus* *49*, 50
Kauai, Hawaii *37*
Kea *Nestor notabilis* *49*, 50
Kettlewell, Bernard 6
keystone predation 29
Koala *Phascolarctos cinereus* 35
Koh Lian Pin 74
Komodo Dragon *Varanus komodoensis* 50
Kona reefs, Hawaii 30
Krakatoa eruption 1883 40–1

Lack, David 44
Lake Baikal, Siberia 17
lakes 17
lampreys 14, *15*
land animals 55
Large Ground Finch *Geospiza magnirostris* 24, *24*
latitudinal gradient 60
length of generations 61
lethal competitors 25–9
Lilford's Wall Lizard *Podarcis lilfordi* 50
limpets (*Patella*) 28
lineage species concept 12–13
Lister, Bradford 44
local abundance 34
Lord Howe Island 16, *18*, 19
Lottia alveus 74

MacArthur, Robert 25, *26*, 38–9, *39*, 41, 42, 45
macroevolution 46, 62
Madagascar 16
Madagascar Periwinkle *Catharanthus roseus* 75
Majerus, Michael 7
Makah Bay, Washington State 29
Malthus, Thomas 4
mammals 50–1, 66
mangroves 39–40
Marine Iguana *Amblyrhynchus cristatus* *49*, 50
marine invertebrates 51, 57
Marshall, Charles 52, *52*
marsupials 35, 68
mass extinctions 50, 54–5, 77, 78
medicines 75
Mediterranean basin **22**, 31, 69
Medium Ground Finch *Geospiza fortis* 23–5, *24*
melanism 6
microbial biodiversity ix
mid-domain effect 60
Mistaken Point, Newfoundland 53, *54*
Moniliformopses (horsetails and ferns) 55, **64**
Monodonta snails 28
Morocco 42–3
morphological species concept 12–13

Moulton, Michael 32
Mount Teide Buglosses *Echium wildpretii 37*
mountains 17, 59
Mundo River Butterwort *Pinguicula mundi 47*
Munroe, Eugene 38
mussels 29
myxomatosis 71

n-dimensional hypervolume 23
Nanmwarki Mwanenihsed 5
narrow niches *47*, 48
natural selection 4–10, *6*, 12–16, 13, 14, 23–5
see also evolution
nature and human enterprise 76
nautiloids 56
near islands 41–2
Nematoda (roundworms) **64**
Nematomorpha (horsehair worms) **64**
New Guinea 41
niche differentiation 23–5
niche segregation 23, 25
see also ecological niches
Northern Pacific Sea Star *Asterias amurensis* 71
Norway 60
Nucleariida (protists) **64**

On the Origin of Species (Darwin) 2–4, 12, 46, 57
Onychophora (velvet worms) **64**
Ordovician period 55

Paine, Robert 29
palm trees *18*, 19
Pantolambda 50, *52*
parallel speciation 13–15
Parma Wallaby *Notamacropus parma* 35
Partula (tree snails) 71
Passenger Pigeon *Ectopistes migratorius*, 68, 72
Passeriformes 36
pebbles 30
penguins 58
Peppered Moth *Biston betularia* 5–7, *6*

Phaeophyceae (brown algae) **64**
Phillips, John 56
Phoronida (horseshoe worms) **64**
phylogenetic species concept 12–13
Pianka, Eric 22
Pimm, Stuart 32, 70
Pingelap Atoll, Micronesia 5
plants 31, 55, 66
Pleistocene megafauna 68
Polar Bear *Ursus maritimus* 73, *73*
polyploids 19–21
Preston, Frank 33
Project XXI (UN) 76–7
Providence, Rhode Island 4–5
Przewalski's Horse *Equus ferus przewalskii* 76
Purple Sea Star *Pisaster ochraceus* 29

Quental, Tiago 52, *52*

rainforests 30–1, *31*
Rambla de Valdemiedes, Murero, Spain 51
rangeomorphs 53
Red Bird of Paradise *Paradisaea rubra* 8
Red Queen hypothesis 53
Red Queen's board 34, 36, 57
regional accumulation 58–9
Rennell Island 41
Repenomamus 50
reptilian dominance 50
'rescue effect' 41
reserves 69
Reznick, David 9–10
Ricklefs, Robert 43
The River of Light (Church) 59
Rock Sparrow *Petronia petronia* 42–3
rockroses (*Cistus* species) 31
Rosemary *Salvia rosmarinus* 31
Rosy Wolfsnail *Euglandina rosea* 70–1
Rough Periwinkle *Littorina saxatilis* 15

satellite species 14
Schluter, Dolph 14

Sclater, Philip 35, 36
Sea Lamprey *Petromyzon marinus* 15
Seilacher, Adolf 53
Semibalanus balanoides 27–9
Seram Island 41
Setophaga (warblers) 25
sexual selection *8, 9,* 61
shallow marine communities 55
Sharp-beaked Ground-Finch *Geospiza difficilis* 48–50
Sierra Nevada, Spain 17
Simberloff, Daniel 39–40
Society Islands 41
solar photovoltaic panels 77
Solomon Islands Skink *Corucia zebrata* 50
Sousa, Wayne 30
South American tropics 59, *59*
Southern House Mosquito *Culex quinquefasciatus* 70
Spain 42–3
Spanish Sparrow *Passer hispaniolensis* 21, 43
specialisation 25
speciation 16, 19, *47*, 48, 50–2
see also ecological speciation
speciation-extinction equilibrium *47*
speciation rate 62, *63*
species 11–16
species–area relationships 33–6, *33*, 42
species equilibrium 38–9, *39*
species massacres 70
species-poor habitats 60–1
Spermatophyta (seed plants) **64**
Stebbins, George 59
stick-insects 15–16
sticklebacks 13–14, 23
storms 29–31, *31*
'struggle for existence' (Darwin) 2–4
The Struggle for Existence (Gause) 25–7
Surtsey, Iceland 40
sustainable development 75–6
sympatric speciation 19

Tawaki Penguin *Eudyptes pachyrhynchus* 50
Tenerife, Canary Islands 37
Terborgh, John 61
threatened species 66–9
Three-spined Stickleback *Gasterosteus aculeatus* 13–14, *13*
Through the Looking-Glass (Carroll) 53
Thylacine *Thylacinus cynocephalus* 68, 70
thyme (*Thymus* species) 31
Timema 15–16
Tocororo *Priotelus temnurus* 43
trap-22 effects 61
tree snails 71
tribal inbreeding 5
trilobites 51, *52*
Trinidad 9
tropical rainforests 31
tropical speciation and extinction 59
tropics 59–60, *59*
Tutt, James William 5–6

underwater volcanos 38
unoccupied niches 50
Urochordata (ascidians, tunicates) **64**

Van Valen, Leigh 53
vascular plants 55
'vendobionts' (Seilacher) 53
Verneuil, Édouard de 51

Wallace, Alfred Russel 35, 36
Wallace's regions 34–6, *35*
warblers 25
water fleas (amphipods) 17
West Indies 43–4
wild grasses 27
Williamson, Mark 72
Wilson, Edward 38–40, *39*, 41, 42, 45
woodpeckers 48

Xanthophyceae (yellow-green algae) **64**

Yellow-rumped Warbler *Setophaga coronata* 26